CLIMATIC AND ENVIRONMENTAL THREATS TO CULTURAL HERITAGE

Climatic and Environmental Threats to Cultural Heritage examines the challenges that environmental change, both sudden and long-term, poses to the preservation of cultural material.

Acknowledging the diversity of human cultural heritage across collecting institutions, heritage sites and communities, the book highlights how, in Australia, Southeast Asia and the Pacific, the quest to preserve such precious knowledge relies on records and narratives being available to inform decisions now and into the future. Bringing together a diverse range of stakeholders who have an interest in – and responsibility for – the care of cultural heritage material and places of cultural heritage value, the book explores their thinking on and actions in relation to issues of climate change and environmental risk. Sloggett and Scott highlight the stakeholders' shared interest in drawing on their expertise to meet the challenges that environmental change brings to the future of our cultural heritage and our cultural identity. Based on the understanding that this global challenge requires local, national and international co-operation, the book also considers how local knowledge can have international application.

Climatic and Environmental Threats to Cultural Heritage will be of interest to those engaged in the study of heritage, conservation, archaeology, archives, anthropology, climate change and the environment. It will also be useful to practitioners and others attempting to understand the effect of environmental change on cultural heritage around the globe.

Robyn Sloggett is an internationally recognised expert in Cultural Materials Conservation. She holds the Cripps Foundations Chair of Cultural Materials Conservation and is the Director of the Grimwade Centre for Cultural Materials Conservation. She has qualifications in Art History, Philosophy and Cultural Materials Conservation. Her work has been recognised by a number of national awards including as a Member of the Order of Australia. In 2021, she was awarded the Marles Medal from the University of Melbourne, which recognises excellence in research impact and was awarded for 'interdisciplinary, cross-cultural research in arts conservation and the significant impact it has had on the communities with who she has collaborated'.

Dr Marcelle Scott is a Research Fellow at the University of Melbourne's Grimwade Centre for the Conservation of Cultural Material where her research focuses on conservation theory, ethics and pedagogy. She draws on her extensive industry experience to research, develop and deliver interdisciplinary and people-led models of cultural materials conservation. Her current research focuses on work-integrated learning and community-based conservation partnerships and programs. She is a recipient of the AICCM Medal, Professional Member of the AICCM and a fellow of the International Institute for Conservation of Historic and Artistic Works (IIC).

CLIMATIC AND ENVIRONMENTAL THREATS TO CULTURAL HERITAGE

Robyn Sloggett and Marcelle Scott

Routledge
Taylor & Francis Group
LONDON AND NEW YORK

Cover image: Sarah McConnell, Mt Ruark 2018. Etching, aquatint, sugarlift. 36 × 26 cm. © The Artist

First published 2023
by Routledge
4 Park Square, Milton Park, Abingdon, Oxon OX14 4RN

and by Routledge
605 Third Avenue, New York, NY 10158

Routledge is an imprint of the Taylor & Francis Group, an informa business

British Library Cataloguing-in-Publication Data
A catalogue record for this book is available from the British Library

ISBN: 978-0-367-75638-3 (hbk)
ISBN: 978-0-367-75636-9 (pbk)
ISBN: 978-1-003-16331-2 (ebk)

DOI: 10.4324/9781003163312

Typeset in Bembo
by MPS Limited, Dehradun

CONTENTS

FIGURES

ACKNOWLEDGEMENTS

The authors are indebted to the many people who supported and contributed to this book.

To Hannah Stewart, a wonderful writer in her own right, who for this book extended her role of Research Assistant to one of friend and colleague, who found the hard to find references and filed them in such a sensible way that even we could find them again, who gently pointed out and corrected errors and ensured we had everything we needed when it was time to submit the final manuscript, thank you.

To the wonderful editors and staff at Routledge, especially Commissioning Editor Heidi Lowther for the initial invitation to write this book, her enthusiasm for the topic, helpful and considered insights, and gentle nudges along the journey, immense thanks. Thanks also to Emmie Shand, Kangan Gupta, and Manus Roy for steering and supporting the project with patience and good counsel. Thank you particularly to Sumit Kumar who rigorously checked each word to ensure that the final both made sense and was properly presented.

This book is about the lived experience of so many people who are working to preserve their heritage in face of escalating challenges wrought by climate change. We are extremely grateful to all those who took the time to respond to our surveys, their thoughtful, informed, and carefully considered insights shaped the way we thought, and wrote, about the topics covered in this book.

We appreciate permission to use the images from Australia ICOMOS; ICCROM and the Canadian Conservation Institute; Shirley Purdie, Warmun Art Centre; Maxime Aubert, Griffith University; Mark Moore and Steven Ahoy, University of New England; The Climate Council; and Bruce Miller, which added rich detail to the story we could tell.

PREFACE

Robyn Sloggett and Marcelle Scott

Between the time this book was conceived and the time it was written, the Covid pandemic had gripped the world, and nations, governments, public institutions, corporations, families and individuals were meeting the challenge with rapid and generally congruent responses. Planes stopped flying, cars were garaged, container ships were docked, greenhouse gas emissions declined and the planet took a big deep breath. (Seemingly the perfect time to lock down and write a book.)

This is in stark contrast to the global responses to the climate crisis; responses that are characterised in many instances by government recalcitrance, incremental and slow reactions, and policies that are not only unhelpful in stemming climate change but that signal a denial that any response is necessary, or worse excuse their inaction by labelling practical action as damaging (to economies, to lifestyles, to income and more). Sometimes the politics of climate change denial claims heritage in defence of doing nothing, arguing that attempts to reduce greenhouse gas emissions will threaten ways of life that have built community identity over generations.

As news came each day of increased loss of life in the struggle against the SARS-CoV-2 virus, news relating to the global climate crisis similarly escalated in severity from one event to the next. Records that were said to demonstrate a 'once in a hundred-year event' were broken in 2020, and again in 2021, and again as we entered 2022. Newspapers led with stories of 'the worst fires' or 'the most devastating floods'. At the same time, in the official Press Release that accompanied the launch of the Intergovernmental Panel on Climate Change's Sixth Assessment Report, Hans-Otto Pörtner, co-chair of the IPCC's Working Group II, concluded that 'Any further delay in concerted global action will miss a brief and rapidly closing window to secure a liveable future'.

This book is an attempt to consider the ways in which heritage, in all its rich and diverse forms, intersects with the catastrophic climate crisis that is facing our planet and to which many people appear disinclined to respond. The reasons for two conservators to write about climate change are relatively straightforward (apart from the fact that conservators are often called upon to explain why their work is not about reforestation, or saving species from extinction). Conservation is about making sure that the past can be taken forward into the future. Heritage is part of the ecosystem of human knowledge that is both a legacy from the past and a resource for the future. As the Australian Institute for the Conservation of Cultural Materials (AICCM) Code of Ethics identifies, conservators work 'to care for cultural material', 'prevent loss of information' and 'enhance its interpretation'.

Nevertheless, exploring the relationship of heritage to climate change is a complex proposition. There is so much at risk in the current climate crisis, and heritage, in all its iterations, is no exception. Simply identifying these risks could easily provide the basis for a number of books. Similarly, assessing the response of the various heritage professions to this crisis, including practical and philosophical responses from archaeologists, archivists, curators, heritage architects, historians and so on, could provide enough material for a series of volumes. The topic also compels analysis of the broader ecosystem in which heritage professionals operate; of people who care for heritage, personally in their families and in their communities, of heritage experts and senior knowledge holders who hold cultural and ecological knowledge of specific communities and areas, and of governments and organisations that provide support. Finally, there is much scope to examine the proposition that heritage is valuable in the climate crisis precisely because it can be used to understand the past, investigate the pathways from that past to the present and create new solutions leading into the future.

However, the lessons of history and the significance of heritage can appear insubstantial in the face of rising sea levels that turn entire nations into migrant populations, when one in one-hundred-year fire and flood events are devastating communities on an annual basis, and when the call to arms in each intergovernmental report becomes more and more urgent and less perceptibly effective. The danger that heritage may appear inconsequential is real. In the face of an impending world catastrophe, the rise of Nazism, philosopher Karl Popper, decried the idea that studying the past could secure predictions of the future, arguing that 'the future relies on us'.[1] Popper did, however, provide a critical analysis of historicism as a tool for human agency for the future, as part of the development of knowledge which makes it true that '*we cannot anticipate today what we shall know only tomorrow*' (original italics).[2]

This is why heritage is more than history. Heritage is culture, memory, activity and above all a community's belief in the value of the past for the future. Today, as the planet faces a crisis of existential proportions, more than any other time in the past the future relies absolutely on the decisions made by current generations. We may not be able to anticipate what we might know, or have to deal with in the future, but it is heritage that provides a strong framework for understanding what knowledge we can draw on as we seek to develop new knowledge and find new ways of acquiring knowledge to deal with the climate crisis. There is no doubt that the climate crisis presents challenges that are so overwhelming, with solutions so seemingly complex, that the world is gripped by a kind of mass catatonia, characterised by inertia, inaction, inattentiveness and, in many cases, a sense of overwhelming grief. In the face of this immense challenge, and armed with the greatest access to the repositories of human knowledge of any generation, a new interest in, and use of heritage, is emerging. After all, it is history that enables us to understand how we got here, and it is culture that will help to determine how we respond. Heritage is both of these. It is this complexity that propelled the writing of this book.

Addressing the impacts of climate change within the heritage sector will require dedication. This is more than establishing committees, strategies, protocols and compiling guidelines for actions. It means engaging with the very real trauma of loss and grief, and of providing counsel and support to people whose lives and sense of identity are under threat. It means recalibrating professional codes to incorporate the kinds of flexible and responsive actions needed in a time of crisis. It means creating programs that are effectively situated, socially and politically.

It was these views, challenges and discussions that compelled us to begin the research that would ultimately lead to this book. We began by asking ourselves the questions:

> What is at risk and why? and,
>
> Where can heritage experts be best positioned to defend the cultural heritage that defines the identity, and supports the well-being, of humankind?

Each chapter in this book explores these questions and many of the responses to them. While we hope the book is useful, and at times enjoyable and at others confronting, we know the answers will only come through ongoing, committed and effective dialogue and action. This book remains, above all, a plea for those engaged in caring for heritage to become active in the struggle to mitigate the climate disaster we are currently facing.

Notes

1 Popper, K 1945, *The Open Society and Its Enemies*, Routledge, UK, p. xxxvii.
2 Popper, K 1957, *The Poverty of Historicism,* Routledge and Kegan Paul, UK, p. xii.

References

Australian Institute for the Conservation of Cultural Materials (AICCM) 2002, *Code of Ethics for the Practice of Cultural Materials Conservation in Australia.*

Popper, K 1945, *The Open Society and Its Enemies*, Routledge, UK.

Popper, K 1957, *The Poverty of Historicism*, Routledge and Kegan Paul, UK.

INTRODUCTION

Robyn Sloggett and Marcelle Scott

For all the reasons that professionals engaged in preserving heritage need to think about climate change, two are particularly compelling. One is that heritage, in all its manifestations, is the receptacle for particular forms of knowledge and memory that have a critical and unique part to play in responses to the challenges of climate crises. The second is that as risks to cultural heritage increase commensurate with the scale of the climate crisis everyone needs to have a voice and to be prepared to act. Heritage experts, and the term is necessarily used broadly, are well-placed to take a lead in both activating public reaction and demonstrating effective responses to climate change. Exploring the role of heritage within the challenges of the climate crisis and examining the contribution made by those with a responsibility to preserve heritage are the foci of this book.

The book sits within the context of existing literature in the field, including historical overviews (e.g., JR McNeill's 2001 edited volume *Environmental History in the Pacific World* and PF Smith's 2014 book *Climate Change and Cultural Heritage: A Race against Time*); political assessments (such as Nowrin Tabassum's recent publication *The Politics of Climate Change Knowledge*); sustainability (CK Hofbauer, EM Kandjani and JMC Meuwissen's 2018 edited work *Climate Change and Sustainable Heritage*); and management (a key example being C Sabbioni, P Brimblecombe and M Cassar's edited volume *The Atlas of Climate Change Impact on European Cultural Heritage: Scientific Analysis and Management Strategies* from 2012). There is also a range of more subject-specific texts such as *Public Archaeology and Climate Change* (2017) edited by T Dawson, C Nimura, E López-Romero and M-Y Daire; and M Hassan and H Xie's book *Climate Change and Conservation of Coastal Built Heritage* (2020).

Nevertheless, the magnitude of the challenge means that much remains to be written on the subject. In 2017, Sandra Fatorić and Erin Seekamp conducted a literature review, responding to the concern that 'a comprehensive

DOI: 10.4324/9781003163312-1

understanding of the impacts of climate change on cultural heritage and resources across various continents and disciplines is noticeably absent from the literature'.[1] They concluded that 'Climate change impacts on tangible and intangible heritage have not received much attention despite broad recognition by UNESCO'[2] and that greater engagement with communities is necessary given that 'values are at the crux of cultural heritage and resources management decision given the embedded place meanings held by an array of stakeholders'.[3] They noted that in over half the publications they surveyed 'three [disciplines] are dominant – architecture and built environment, climate and natural hazards, and archaeology'.[4] They also identified additional research that needed to be done to support the application of 'more of a bottom-up approach to preservation and adaptation of cultural heritage and resources' suggesting that more work using 'qualitative participatory-based methods' was needed.[5] Finally, they pointed out that research in the field was 'geographically limited', with a predominant focus on Europe and North America, and that '[v]ery limited research has been undertaken in Africa, Asia and South America'.[6]

Those engaged in the care of cultural heritage are very familiar with managing risk. It is a concern that sits at the core of heritage conservation, alongside concepts of 'safeguarding' and 'protecting'. The need to safeguard built heritage and sites is the focus of Article 2 and Article 3 of Definitions in the 1964 ICOMOS Venice Charter, which declares that 'The conservation and restoration of monuments must have recourse to all the sciences and techniques which can contribute to the study and safeguarding of the architectural heritage'; and that 'The intention in conserving and restoring monuments is to safeguard them no less as works of art than as historical evidence'.[7] In cultural materials conservation, the definition of the term *conservation* was adopted by the International Council of Museums-Committee for Conservation (ICOM-CC) in New Delhi in 2008, as 'all measures and actions aimed at safeguarding tangible cultural heritage while ensuring its accessibility to present and future generations'. This definition continues, 'conservation embraces preventive conservation, remedial conservation and restoration. All measures and actions should respect the significance and the physical properties of the cultural heritage item'.[8]

These are definitions about what to do. Why it is important to act is a question that was addressed in The Burra Charter (2013), as an advocacy statement that positions cultural heritage as central to human identity:

> Places of cultural significance enrich people's lives, often providing a deep and inspirational sense of connection to community and landscape, to the past and to lived experiences. They are historical records, that are important expressions of Australian identity and experience. Places of cultural significance reflect the diversity of our communities, telling us about who we are and the past that has formed us and the Australian landscape. They are irreplaceable and precious.

> These places of cultural significance must be conserved for present and future generations in accordance with the principle of inter-generational equity.
>
> The Burra Charter advocates a cautious approach to change: do as much as necessary to care for the place and to make it useable, but otherwise change it as little as possible so that its cultural significance is retained.[9]

In a world where change is constant, unremitting and blisteringly fast there is a deep comfort in the suggestion that to change as little as possible is the best approach; and that to intervene as little as possible gives us the best chance of retaining significance. Unfortunately, the world's heritage, like everything else on the planet, no longer has the luxury of slow, minimal responses. The climate crisis has catapulted every industry, every profession, every discipline, every community and every person into the need to respond immediately, or preferably, yesterday. In the past, risk management and conservation responses were examined within the framework of the 'ten agents of deterioration', identified as physical forces, fire, pests, light, incorrect relative humidity, thieves and vandals, water, pollutants, incorrect temperature and dissociation.[10] Disasters sat as singular examples that required singular, albeit often complex, responses. In the second decade of the twenty-first century, the impact of climate change has meant that cataclysmic events often bring all ten agents of deterioration together in one devastating incident, or equally as probable, that all ten agents are part of slow shift in the overall environment in which heritage exists. Rather than respond directly to the need to manage risk, more comprehensive approaches that take into account resilience, mitigation and adaption play a significant role in climate crisis-induced risk management.

These are professional responses however, and while climate change is a global concern, this book is principally concerned with heritage and the effects of, and responses to, the climate crisis in Southeast Asia and the Pacific, as part of the broader area which has more recently attained political regional status as the Indo-Pacific region. With a focus on this region, there was a need to broaden the nomenclature that describes heritage and those who care for and manage it. This was prompted by the generality and incongruity in the practical, philosophical and logistical use of the terms 'heritage' and 'heritage professional'. UNESCO's 1972 *Convention Concerning the Protection of the World Cultural and Natural Heritage* separates heritage into 'cultural heritage', constituted by 'monuments', 'groups of buildings' and 'sites', and 'natural heritage' comprising 'natural features', 'geological and physiographical formations' and 'natural sites'.[11]

Faced with the need to more fully account for the broad range of the ways in which human cultural activity has meaning within the built, collected and natural environments UNESCO developed another Convention, the 2003 *Convention for the Safeguarding of the Intangible Cultural Heritage*. This created another category of heritage, which is described as

> the practices, representations, expressions, knowledge, skills – as well as the instruments, objects, artefacts and cultural spaces associated therewith – that communities, groups and, in some cases, individuals recognize as part of their cultural heritage. … and is manifested inter alia in the following domains:
>
> a. Oral traditions and expressions, including language as a vehicle of the intangible cultural heritage;
> b. Performing arts;
> c. Social practices, rituals and festive events;
> d. Knowledge and practices concerning nature and the universe;
> e. Traditional craftsmanship.[12]

For many societies synchronous views of human existence and the natural environment mean that the separation of 'cultural' and 'natural' heritage from 'intangible' heritage is a false dichotomy. For this reason, this book takes a broader but more integrated view of heritage, and for this purpose South Africa's National Heritage Council's current working definition of heritage provides a useful framework:

> Heritage is what is preserved from the past as the living collective memory of a people not only to inform the present about the past but also to equip successive generations to fashion their future. It is what creates a sense of identity and assures rootedness and continuity, so that what is brought out by dynamism of culture is not changed for its own sake, but it is a result of people's conscious choice to create a better life.[13]

For members of many societies, across very diverse communities, the idea that they are preserving a dynamic heritage is the predominant way of thinking about heritage preservation, and presupposes collaborative, communal and cooperative ways of sharing cultural heritage and keeping knowledge alive. This is somewhat removed from the concept of the heritage professional who attains qualifications, produces written documentation based on rigorous professional standards and translates community ideas of value and significance into public policy and government decision-making. For this reason, the term 'experts' rather than 'professional' is used throughout this book to better reflect this combination of interrelated practices and people. In acknowledging this diversity of cultural expertise, this book privileges the use of the term 'heritage knowledge experts', as a broader category within which the term 'heritage professional' is located as a subcategory. The rationale for this is twofold. First, knowledge about cultural heritage in both tangible and intangible forms is shared in communities as well as held by senior knowledge holders. Most of these knowledge holders, while global experts in their knowledge domain, would not be defined as 'heritage professionals' and significant iterations of heritage would be excluded if their knowledge and the ways it is nurtured and transmitted were not able to be considered. A much more socially and culturally pluralistic, historically valid and philosophically robust definition was required. Nevertheless, there is a particular role for the heritage professional in debates around and responses to climate change, so the term is retained for specific uses in this book.

Another problem of categorisation terminology rests in the various types of sub-categories of heritage professionals that exist. Conservators, curators, librarians, archivists, collection managers, archaeologists, heritage architects, historians, linguists and others are all heritage professions engaged in cultural conservation and heritage preservation. By necessity, because of the nature of the relationship of heritage to climate change, all are included in some form, at some point, in this book. All have specific professional bodies and societies which represent them, and which deliver annual conferences, publications and other forms of public exchange where knowledge is shared, contested and developed. They also have clear codes of ethics and codes of practice or guidance, and practice notes, that provide frameworks for research, decision-making and practical responses of the kind needed to address the risks presented by climate change. All employ assessment frameworks as a standard tool in decision-making.[14] The function of standard frameworks enables effective mitigating actions to be put in place. The type of response will, in turn, depend on the context in which the risk is emerging. This provides secure standards within which to locate professional activity, while at the same time enabling structured, and agreed parameters for the type of innovation needed to respond to various climate change issues. In many communities, senior knowledge holders and cultural leaders provide similar guidance, so a nuanced understanding of terminology is important.

A similar need arose from the generality and incongruity in the practical, philosophical and logistical use of the terms 'climate change' and 'climate crisis' and other associated terminology. In October 2019, *The Guardian* updated its editorial style guide 'to introduce terms that more accurately describe the environmental crises facing the world'.[15] These included:

> 'Climate emergency' or 'climate crisis' rather than 'climate change'
>
> 'Climate science denier' or 'climate denier' in place of 'climate sceptic'
>
> 'Global heating' not 'global warming'
>
> 'Greenhouse gas emissions' in preference to 'carbon emissions' or 'carbon dioxide emissions'
>
> 'Wildlife' not 'biodiversity'
>
> 'Fish populations' not 'fish stocks'.

This book takes a similar approach by attempting, where possible, to use terms that are precise, credible and unambiguous. Where the term 'climate change' is used, it bears the full weight of the connotations of crisis and disaster identified earlier.

In a commentary piece in *Nature Climate Change* in 2016, political geographer Jon Barnett and colleagues argue that the threat of climate change requires 'socially engaged research that explains what people value highly, how climate change imperils these phenomena, and strategies for embracing and managing grief' so that loss can be better prepared for, risks better understood and future suffering reduced.[16] This book attempts to respond to this challenge through a multi-disciplinary analysis of key issues, and an assessment of various interdisciplinary methodologies that appear to have potential as effective tools in considering the preservation of heritage in climate crises and emergencies.

Chapter 1 explores the links between heritage and story-telling with places, objects and performance providing the point of departure for recounting memories, experience, lore and lessons. Exploring Narrative Policy Framework Theory, this chapter demonstrates how attempts to build effective risk management and resilience strategies are supported by the ways in which narratives develop. Narratives feed knowledge across professional and societal divides, and so are useful when engaging with the plurality of voices needed to address risks to heritage posed by the climate emergency. Heritage experts are particularly effective users of narrative frameworks as part of significance assessment, and risk management documentation.

Chapter 2 examines the question: What do records tell us about climate change? It describes how citizen science projects are compiling new knowledge from old records, and how histories that have remained hidden for a century or more are being used in the fight against climate change. The chapter highlights how an alignment between heritage organisations, sectors and communities can support new enquiries into climate change and its impacts.

Chapter 3 considers accounts from Australian Indigenous communities, whose history encompasses extreme climate change, most spectacularly and recently, the last Ice Age that occurred some 11,000 years ago. It also considers the impacts of language barriers to show language can be used, unthinkingly to exclude senior Indigenous knowledge holders, and therefore much Indigenous knowledge, from high-level discussions on climate change impacts and response.

Chapter 4 examines how, as significant repositories of cultural knowledge, rock art sites are at risk. These sites are the record of Indigenous people who lived through some of the most cataclysmic climate change events in human history, including the last glacial period some 25,000 years ago. They provide evidence of species, now extinct, and of the changes in climate over millennia. With rising sea levels, coastal rock art sites are at risk from incremental damage inflicted by slow-onset emergencies such as gradual rises in sea level as well as rapid-onset disasters including bushfires and floods.

Chapter 5 presents case studies relating to natural disasters from Timorese diaspora in Australia. It explores how the retelling of memories and histories produces significant resources that can be utilised to benefit the process of nation-building, and help to provide a continued sense of identity in the midst of disruption and change. This chapter demonstrates the value of memory as intangible heritage in diasporic communities.

Chapter 6 draws on personal accounts from recent devastating Australian bushfires and the shifting debate around the impacts of climate change on fire intensity and behaviour. It illustrates how, as the 'Climate Change Debate' rages across political, social and environmental networks, loss remains profoundly personal. The second part of the chapter describes the ways in which traditional knowledge is being used to mitigate the effects of climate change and is being employed to 'keep country strong' and protect cultural heritage sites.

Chapter 7 provides an overview of the types of risks that are faced by heritage places and collections and their custodians in the Indo-Asia-Pacific region. It examines examples of both slow-onset and rapid-onset disasters to illustrate why cultural heritage, both tangible and intangible, is important in understanding, mitigating and responding to such disasters. It also considers how personal accounts ground statistics to produce a better understanding of impacts and responses to climate change. In doing so, it brings together the meta-narrative of UNESCO and other globally positioned organisations with the local narratives of individuals' lived experiences.

Chapter 8 provides an overview of the responses to risks from climate change by professional bodies, institutions and heritage organisations that represent the disciplines and professions working across cultural heritage conservation and climate crisis mitigation. It presents a selected chronological account of key statements and developments across the sector, identifies specific initiatives, the form that this work has taken and the views of professionals and organisations in regard to such initiatives.

Chapter 9 draws on the views of conservation educators and students to explore how the lens of climate change and environmental risk can be focused within a conservation curriculum to assist students realise their professional responsibility to be agents of change in this critical and emerging sphere of conservation practice.

Political geographer Jon Barnett and colleagues argue for a science of loss that brings together 'insights and methods from critical heritage studies, archaeology and museum studies', all areas where memorialising of loss and people's relationships with place are examined, with approaches that anticipate future loss.[17] This book is an attempt to go some way to do this, but there is much more that could be written and in many ways this book is, at its core, a series of sign-postings to where considerably more research could be done, and many, many more books could productively follow.

Notes

1 Fatorić, S & Seekamp, E 2017, 'Are cultural heritage and resources threatened by climate change? A systematic literature review', *Climatic Change*, Vol. 142, No. 1–2, p. 228.
2 Fatorić & Seekamp 2017, p. 235.
3 Fatorić & Seekamp 2017, p. 241.
4 Fatorić & Seekamp 2017, pp. 237–8.
5 Fatorić & Seekamp 2017, p. 240.
6 Fatorić & Seekamp 2017, p. 240.

7 ICOMOS 1964, *International Charter for the Conservation and Restoration of Monuments and Sites (The Venice Charter 1964)*, articles 2–3.
8 ICOM-CC 2008, *Terminology to Characterize the Conservation of Tangible Cultural Heritage*, ICOM-CC, accessed on 13 Oct 2021, https://icom-cc.org/en/terminology-for-conservation
9 Australia ICOMOS 2013, *The Burra Charter: The Australia ICOMOS Charter for Places of Cultural Significance*, p. 1.
10 Canadian Conservation Institute (CCI) 2017, *Agents of Deterioration*, accessed on 13 October 2021, https://www.canada.ca/en/conservation-institute/services/agents-deterioration.html
11 UNESCO 1972, 'I. Definition of the cultural and natural heritage: article 1 and article 2', *Convention Concerning the Protection of the World Cultural and Natural Heritage.*
12 UNESCO 2003, 'I. General provisions, article 2 definitions', *Convention for the Safeguarding of the Intangible Cultural Heritage.*
13 National Heritage Council South Africa 2020, *2020/2021–2024/2025 Medium Term Strategic Framework*, National Heritage Council, Pretoria, South Africa, p. 8.
14 See Australian Institute for the Conservation of Cultural Materials (AICCM) 2002, *Code of Ethics for the Practice of Cultural Materials Conservation in Australia.*
15 Zeldin-O'Neill, S 2019, "It's a crisis, not a change': the six Guardian language changes on climate matters', *The Guardian*, 16 October 2019.
16 Barnett, J, Tschakert, P, Head, L & Adger, WN 2016, 'Commentary: a science of loss', *Nature Climate Change*, Vol. 6, p. 976.
17 Barnett et al 2016, p. 978.

References

Australia ICOMOS 2013, *The Burra Charter: The Australia ICOMOS Charter for Places of Cultural Significance*, Australia ICOMOS Incorporated, Burwood.

Australian Institute for the Conservation of Cultural Materials (AICCM) 2002, *Code of Ethics for the Practice of Cultural Materials Conservation in Australia*, AICCM, Canberra.

Barnett, J, Tschakert, P, Head, L & Adger, WN 2016, 'Commentary: a science of loss', *Nature Climate Change*, Vol. 6, pp. 976–978. doi:10.1038/nclimate3140

Canadian Conservation Institute (CCI) 2017, *Agents of Deterioration*, accessed on 13 October 2021, https://www.canada.ca/en/conservation-institute/services/agents-deterioration.html

Dawson, T, Nimura, C, López-Romero, E & Daire, MY 2017, *Public Archaeology and Climate Change*, Oxbrow Books, Oxford.

Fatorić, S & Seekamp, E 2017, 'Are cultural heritage and resources threatened by climate change? A systematic literature review', *Climatic Change*, Vol. 142, No. 1–2, pp. 227–254.

Hassan, M & Xie, H 2020, *Climate Change and Conservation of Coastal Built Heritage*, Springer, Singapore.

Hofbauer, CK, Kandjani, EM & Meuwissen, JMC (eds) 2018, *Climate Change and Sustainable Heritage*, Cambridge Scholars Publishing, Newcastle upon Tyne.

ICOM-CC 2008, *Terminology to Characterize the Conservation of Tangible Cultural Heritage*, ICOM-CC, accessed on 13 October 2021, https://icom-cc.org/en/terminology-for-conservation

ICOMOS 1964, *International Charter for the Conservation and Restoration of Monuments and Sites (The Venice Charter 1964)*, accessed on 12 July 2022, https://www.icomos.org/charters/venice_e.pdf

McNeill, JR (ed) 2001, *Environmental History in the Pacific World*, Routledge, UK.

National Heritage Council South Africa 2020, *2020/2021–2024/2025 Medium Term Strategic Framework*, National Heritage Council, Pretoria, South Africa.

Sabbioni, C, Brimblecombe, P & Cassar, M (eds) 2012, *The Atlas of Climate Change Impact on European Cultural Heritage: Scientific Analysis and Management Strategies*, Anthem Press, London.

Smith, PF 2014, *Climate Change and Cultural Heritage: A Race against Time*, Routledge, UK.

Tabassum, N 2022, *The Politics of Climate Change Knowledge: Labelling Climate Change-Induced Uprooted People*, Routledge, UK.

UNESCO 1972, 'I. Definition of the cultural and natural heritage: article 1 and article 2', *Convention Concerning the Protection of the World Cultural and Natural Heritage*, accessed on 10 July 2022, https://whc.unesco.org/en/conventiontext/

UNESCO 2003, 'I. General provisions, article 2 definitions', *Convention for the Safeguarding of the Intangible Cultural Heritage*, accessed on 12 July 2022, https://ich.unesco.org/en/convention

Zeldin-O'Neill, S 2019, "It's a crisis, not a change': the six Guardian language changes on climate matters', *The Guardian*, 16 October 2019.

1

THE STORY OF CLIMATE CHANGE: NARRATIVES AS INFLUENCERS

Robyn Sloggett and Marcelle Scott

There are three conditions that are necessary for something to be valued as heritage. First, it needs to have come from the past to the present, accompanied by a desire that it continues into the future. This makes the idea of inheritance a condition of heritage preservation. Second, and linked to this, there needs to be a community that values this possibility. Third, there must be the idea of what risk looks like, and how the value of the heritage might be preserved and protected. This third condition underpins all activities associated with heritage from the intergenerational transfer of knowledge through stories and performance, to the building of national collecting institutions. Because heritage is defined by these three conditions, understanding the significance and value of heritage is predicated on listening to what people think about their heritage and the risks to it. With this in mind, this chapter explores how heritage professionals can best engage with individuals to address the risks posed by climate change. It takes its point of departure from the call made by geographers Jon Barnett, Petra Tschakert, Lesley Head and W. Neil Adger for the development of a 'science of loss' that brings together the insights and methods from the disciplines engaged in heritage studies, with a broader range of interdisciplinary methodologies to assist in developing appropriate action, mitigation and response to events resulting from climate crises.[1] In particular, this chapter examines how stories about loss, and about the potential of loss, can be used to engage and activate people in discussions of heritage and risk from climate change.

Narrative policy framework, as a method in political science and public management studies, helps to construct a systematic approach to the stories people tell and why and how they tell them. This approach addresses the ways in which individuals perceive, make sense of, and respond to crises, and therefore it has the potential to be useful in examining the relationship between heritage conservation and climate change. While individual responses occur at the micro-level, narrative

DOI: 10.4324/9781003163312-2

policy framework theory is also useful because it helps to inform understanding of the meso level at which individuals operate within communities.[2] By beginning with the people who value their heritage, professionals are better able to understand where they might fit in the complex chain of response that demonstrates how and why communities choose to take particular actions. It sets the scene for later chapters that examine specific areas of heritage and specific responses to heritage conservation.

Heritage, loss and resilience

The ways in which heritage supports resilience, optimism, meaning and connectedness is a growing area of critical heritage studies.[3] Harvey and Perry argue that an intrinsic connectedness has developed between the idea of the significance of heritage as a values-based argument and ideas of ''vulnerability' and 'threat'' in climate change.[4] They argue that much of the discourse relating to climate change is 'channelled through notions of heritage' and that 'human apprehension of climate change, itself, has a heritage that must be considered'.[5]

In line with calls for a 'science of loss', Cornelius Holtorf links values, loss and increased climate crisis-induced risk, arguing that within the discourse of heritage values, room needs to be made for supporting the ability of individuals and communities 'to accept loss and transformation', contending that:

> An increased resilience and capacity to deal with transformation and even loss of specific manifestations of cultural heritage can help people adapt to new circumstances and absorb adversity in their own lives too. With this in mind, maybe cultural heritage conservation ought not to celebrate so much the inherent values and timeless qualities of the remains of the past bequeathed to us. Instead this cultural heritage might facilitate our capability of adapting those legacies of the past to changing circumstances today and in the future … as inspirations for the need to embrace uncertainty and the human potential to keep developing over time.[6]

In advocating for a 'science of loss', Barnett, Tschakert, Head and Adger maintain that there are 'the primary goods that are essential for a free and dignified life … [which] are the raison d'être of collective action, legal rights, and the state' and that there are other 'phenomena that people value that are at risk from climate change, but that are overlooked because they cannot be captured by standard metrics'.[7] They locate these 'overlooked' phenomena

> at very fine spatial and temporal scales of experience, for example personal items (such as photographs); material artefacts (such as the graves of ancestors); daily practices (such as gardening); and occupational identities (such as those of farmers and fishers).[8]

Much of what is characterised as heritage sits within these 'spatial and temporal scales of experience'.

Exploring local and Indigenous approaches to managing climate change risks for archaeological sites, Carmichael et al emphasise the value of approaches that are 'locally driven and community owned', noting that such approaches are

> consistent with 'community-based adaptation' approaches, with high levels of local engagement and leadership that can reflect local values, knowledge, and capacities. Such 'bottom-up' approaches are a crucial starting point that can be complemented by 'top-down' approaches (e.g., regional coordination or expert archaeological review of cultural significance assessment) in subsequent stages.[9]

This integration of 'bottom up' and 'top down' approaches is not straightforward and is often complicated and impacted by personal responses to stress, grief and loss. This is borne out in a number of studies of climate change-induced environmental crises that examine chronic reactions, such as the stress of ecological grief, described as grief 'associated with the physical disappearance, degradation and/or death of species, ecosystems and landscapes … driven by climate change'.[10] These studies explore the pervasive sense of threat and loss that accompanies descriptions of environmental changes, and acute reactions that manifest as the result of rapid onset events such as fires, floods, cyclones and severe storms.[11] In a report prepared in 2015 for the American Psychological Association titled *Mental Health and our Changing Planet: Impacts, Implications, and Guidance* Clayton et al, citing the work of Sawitri, Hadiyanto and Hadi argue that:

> People's willingness to support and engage in climate solutions is likely to increase if they can relate them to local experiences or if they see the relevance to their own health and well-being. Additionally, individuals who have higher perceived environmental self-efficacy, or the sense of being able to positively contribute, are more motivated to act on climate solutions.[12]

Clayton et al focused on the mental health impacts of climate change from the perspective of psychology-related professions. They emphasised that practical solutions support action and build 'emotional resiliency',[13] and identified five attributes that help individuals achieve resilience:

1. Build belief in one's own resilience.
2. Foster optimism.
3. Cultivate active coping and self-regulation skills.
4. Maintain practices that help to provide a sense of meaning.
5. Promote connectedness to family, place, culture and community.[14]

The function of cultural heritage in providing individuals and communities with a sense of permanence and connectedness overlaps with ideas of emotional resilience and motivation.

The theoretical and methodological frameworks that link resilience, transformation, adaptation and cultural heritage conservation remain under researched. Nevertheless, on a practical level, heritage professionals provide advice that assists individuals, families and communities to look after what they value. Significance assessment and risk assessment methodologies used in heritage assessment and preservation activities build confidence in decision-making by ensuring adherence to professional standards and skills. These are important professional functions that provide security and stability when communities face uncertain futures.

The ability of heritage studies, from cultural materials conservation through to critical heritage theory, to understand the significance at a personal and local level also helps to address some of the major barriers to comprehending the seriousness of climate crises that Clayton et al noted. These include the perception that climate change is 'global, distant, and difficult to understand', that political affiliations affect people's awareness and belief and that the complexity and fear of climate change make people 'feel uncertain and in denial'.[15] They proposed that learning about the local impacts of climate change could 'make the problem more tangible and a reality'.[16] The practical and grounded nature of heritage conservation, specifically through making joint decisions about real objects and places, has the potential, therefore, to be an important aspect of local and global responses to climate change. But how can existing heritage and conservation methodologies be utilised in these broader responses?

One key is the range of standard methodologies that heritage professionals utilise for risk management and decision-making. Many of these, like the Significance 2.0 methodology, the Burra Charter Process (Fig. 1.1) and the ICCROM/CCI *A Guide to Risk Management in Cultural Heritage* (Fig. 1.2) posit an activity ecosystem in which decision-making is an outcome of a shared process of communication and engagement.[17] Climate change is, however, a multi-faceted problem of such immense proportions and complexity that standard top-down professional-led processes, even those that prioritise communication and consultation, may not be effective or even possible. It is not straightforward to identify a position for professional heritage conservation and preservation frameworks within this complex network of practical considerations involving radical new research, responsive planning and immediate action, that are overlain by the visceral impacts of ideology, fear and belief. Climate change can be a compelling story of an impending Armageddon, or an equally compelling story of a politically crafted deceptive misconstruction. Understanding how to tell the story of the value of heritage within this complex mix is important.

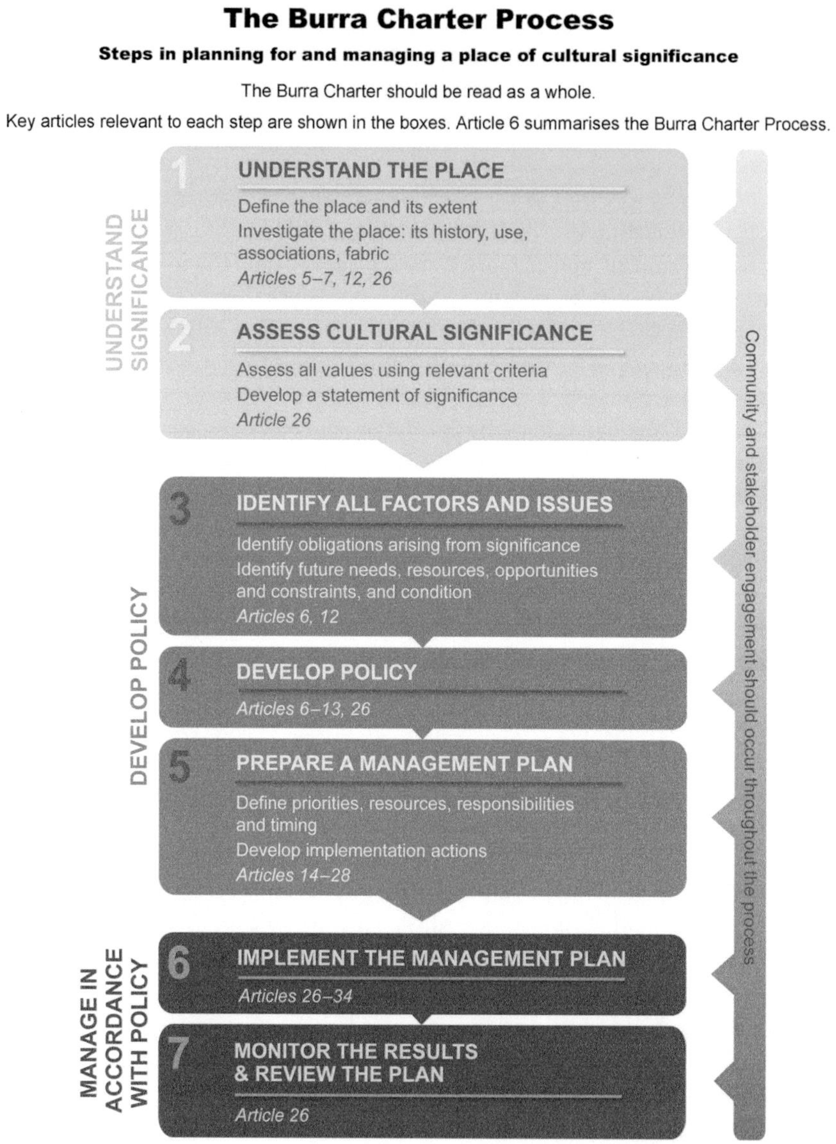

FIGURE 1.1 The Burra Charter Process: flow chart from the *Australia ICOMOS Burra Charter, 2013, p. 10.* © Australia ICOMOS Incorporated 2017.

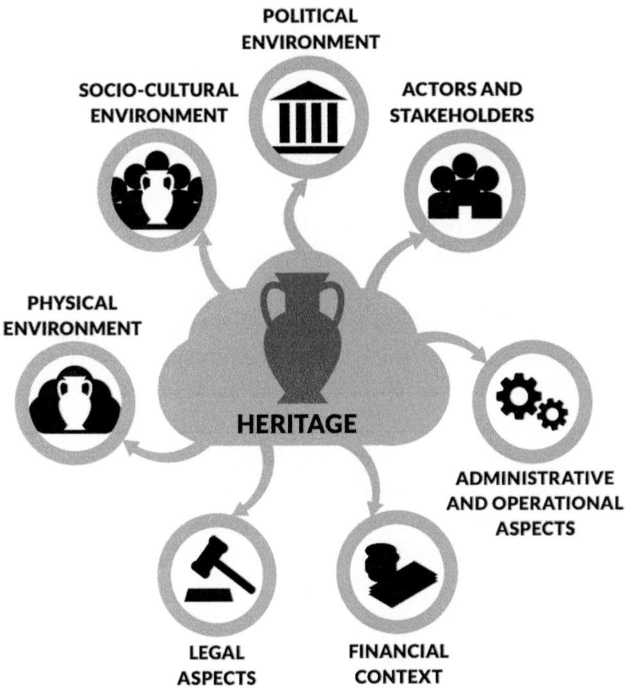

FIGURE 1.2 "Understanding the Context": © ICCROM and Government of Canada, Canadian Conservation Institute

Telling stories

Ask someone in Italy to tell you the meaning of Ash Wednesday and you are most likely to hear an account of the first day of Lent; the day in the Christian calendar that marks 46 days before Easter. On this day of fasting, and celebration, the congregation comes to church to take part in a ceremony where the local priest applies ashes on foreheads of the faithful. The ashes are from the burnt palm fronds that were carried in the church procession on Palm Sunday of the preceding year. The palm fronds symbolise Christ's passage into Jerusalem a week preceding his death and his resurrection. The ashes are a reminder of the mortality of human life, 'ashes to ashes, dust to dust'. Ash Wednesday reminds people of who they are, what they believe in and where they come from.

In Australia, however, if you ask someone to explain the meaning of Ash Wednesday, you are most likely to hear a very different account, where Ash Wednesday is the name of an event that occurred on 16 February 1983 in southeast Australia.

Across the state of Victoria, the day was blisteringly hot. Winds were violent and gusty, picking up the red dust-dry topsoil from the farms of northeast Victoria, blowing tonnes of topsoil across the state to the towns and cities in the southeast. In Melbourne, cars and buildings were engulfed in red dust that covered paths and smothered trees and plants. The temperature rose quickly to 43.2°C (109.7°F), relative humidity dropped to 6%, winds were howling at over 110 km/h (around 68 mp/h). After seven years of drought, the countryside was tinder dry.

In the small mountain township of Cockatoo, some 60-km southeast of Melbourne and with a population of around a thousand around 800 people, people were nervous.[18] Bushfires had started that afternoon in the forests and grasslands about 10 km away at Upper Beaconsfield, and later about 20 km away at Belgrave Heights. By evening, a wind change had diverted the fires away from Cockatoo, and people sat down to their evening meal looking forward to an uneventful, if uneasy, night. Bushfires were a regular occurrence and local townships were well-supported by a network of volunteer firefighters, 'the firies', who were brought together under the umbrella of the Country Fire Authority, otherwise known as the CFA. There was long-term knowledge amongst these communities of how bushfires behaved and fire-fighting knowledge was passed down from one generation to the next. By nightfall, the fire brigade was leading the population of Cockatoo out of the burning township. Not everyone made it. Everyone who survived had a story to tell. Some stories mirrored others, some were quite different. These stories formed the basis for evidence from which recommendations arose. Stories of where people died interlocked with questions of how and why they died. Cause and effect intersected with stories of the buildings and places that had survived, how they had survived, the measures that had been in place to control the undergrowth, the effect that ten years of mainly below average rainfall had had on the bushland, and what actions had been taken to mitigate the risk of fire. Questions arose around the stories of what appeared to be success and what looked like failure, and personal experiences became the data from which to develop new strategies and policies. New sets of guidelines were established. Advice was formulated. Recommendations implemented. Lessons learnt. Australia was ready to tackle the next major bushfire, fore-warned and fore-armed.

On Saturday 7 February 2009, 26 years on from Ash Wednesday, southeastern Australia was once again consumed by wildfires. The toll was even worse; 173 deaths, over 750 homes destroyed, 450,000 hectares burnt beyond recognition. It was called Black Saturday. For days Australians watched as the death toll was revised up, again, and again, and again. Two days later, the Premier of the State of Victoria, John Brumby, announced the establishment of a Royal Commission to investigate 'all aspects of the government's bushfire strategy'.[19] People were angry and eager to give their accounts of what they had lived through. When the four-volume final report was released in July 2010, the three Commissioners made it clear they had listened.

> In our deliberations we ensured that the voices of affected community members were heard. Our priority was to listen to people directly affected by the fires. We also ensured that the Commission's processes were firmly based in the community through open hearings (including in regional towns), web streaming so that people could listen to the hearings over the internet, public submissions, the participation of lay witnesses, the creation

> of the Black Saturday Gallery, and the participation of family and friends in hearings about people who died as a result of the fires.[20]

The Commissioners also saw the report as a cathartic testimonial claiming: 'This report is an important part of securing the memory of the fires'.[21]

The four volumes were written as stories of the event.

> The first volume describes the origins and course of the 15 fires that wrought the greatest harm on 7 February and the response to them. It also tells the stories of the 173 people who died ... Volume IV reproduces the statements of the 100 lay witnesses who gave personal accounts of their experiences in the fires in late January and February and in their aftermath. The stories told by these people grounded our work. They continually reminded all at the Commission that bushfires deeply affect people and communities and that their needs and safety must be at the forefront of government policy.[22]

Despite lessons learned, bushfires continue to grow in extent and severity. The names given to the fires changed to reflect new experiences. No longer defined by a day, fires that burnt continuously during the period September 2019 through to March 2020, became known as Black Summer. Across the country, 18.6 million hectares (around 46,000 million acres) were burnt and new stories were being told about aberrant forms of fire behaviour that had never been seen before. Stories of fires that created their own weather, fires that flipped an eight-tonne fire truck on its side, that destroyed 2799 homes, killed 34 people, and created smoke that blocked views of Sydney Harbour for weeks and was reported on the other side of the Pacific, in Chile and in Argentina. For the first time headline stories included climate change, focussing on the threat of mass species extinction and the impact of tonnes of carbon dioxide emissions.[23] By 2020, climate change was well and truly part of the story of what caused bushfires; and not just in Australia. Similar catastrophic scenarios occurred in Greece and Canada in the following summer. A burning planet became a metaphor for the risks posed by climate change. 'Our house is on fire' became the catchcry for a generation of school children scared of the future and shocked by the inaction of the world's leaders.[24] No longer single events, bushfires in Australia were now part of the broader narrative around seismic disturbances to the natural order of things. The country was gripped with a new kind of national grief, an ecological grief characterised by a sense of loss of intangible values that Cunsolo and Ellis identify as '(amongst others) loss of knowledge, culture, sense of place, identity and social cohesion'.[25]

Narrative frameworks and story-telling

In his discussion of narrative as communication, Walter Fisher explained that:

> From the perspective of the narrative paradigm, the dynamic of this situation is that rival stories are being told. Any story, any form of rhetorical

> communication, not only says something about the world, it also implies an audience, persons who conceive of themselves in very specific ways.[26]

He went on to argue that

> the concepts of public and social knowledge should be reconceived in light of the narrative paradigm … To consider that public-social knowledge is to be found in the stories that we tell one another would enable us to observe not only our differences, but also our commonalities.[27]

The stories we hear and the stories that we tell about climate change are complicated, politicised, polarised, conflicting and at times terrifying. They are about people's livelihoods, the futures they see for their children, and the witness they bear to the loss and destruction of much that is held dear. They describe cataclysmic shifts in our understanding of the world, and of significant change that is, for most people, incomprehensible. The timing of the seasons, the migration of birds, the sound of frogs, the germination of seeds, the curve of a coastline, the rhythms of an ecosystem, these are poetic, nuanced, measured concepts. They are also now part of the more muscular narratives of political decision-making, health, asylum seeking, mental well-being, job security and of the much more abstracted narratives of inheritance, knowledge, responsibility and existence. Narrative Policy Framework theory deals directly with attempts to understand how the translation of stories from individuals, through communities and into the third domain of professional, governance and public policy affects decision-making. In this way, it is a tool for understanding the interrelationships and working attributes that exist between events, research about events, the stories that form the basis for the research on these events, and the policy directives that emerge from this process.

As threats to cultural heritage increase, heritage professionals will need to increase the support available for the interventions needed to mitigate the effects of, or remediate damage resulting from, climate change. If we accept that 'The stories we tell are profoundly influenced by what is possible and what is valued within our culture' then the function of narrative theory in heritage studies is to engage individuals and communities to tell the stories they want to tell.[28] Political scientists, Molly Patterson and Kristen Renwick Monroe, citing earlier work by Somers and Gibson, identify four features of narratives that provide a useful methodological framework: *Ontological narratives*, with 'a connection between identity (understanding of self) and agency (the conditions for action)'; *public narratives*, presenting 'narratives of institutions or social formations'; *conceptual narratives*, constructed by social scientists to examine 'social forces such as path dependency, political institutionalisation, and economic growth; and, finally, *meta-narratives*, that are the 'grand narratives of our time in which we are embedded as social actors and social scientists' and involving 'epic dualities, such as the individual vs society or order vs chaos/anarchy'.[29] The links between personal identity, institutions, politics and economics and the big issues in society reflect the domains in which many heritage professionals operate. Narrative frameworks are used consistently by heritage experts to tell stories, to manage representation and to articulate what is valued from the past and needed for the future.

Patterson and Monroe also point to the fact that 'experience is necessarily mediated by our understanding of the world'.[30] For heritage professionals, understanding the link between multiple narratives, including at times conflicting narratives and final decision-making helps to understand how individuals or communities hold systems of belief that underwrite public positions and decision-making, and affect an individual or community response to various proposals. For example, in the face of rising sea levels, a community might be asked to agree to protect a collection or heritage building by moving it to a safer location; or keeping it in situ and investing in infrastructure such as a seawall; or doing nothing. As professional codes of ethics and practice require that stakeholders have a role in policy decision-making, how heritage experts understand, negotiate and engage with competing narratives can result in being able to protect heritage, or having advice overridden.

In a case study that examined the management of flood risk posed by the vulnerability of the Yellowstone River, Shanahan, Raile, French and McEvoy brought a narrative policy framework to bear on issues relating to 'climatological and meteorological conditions like snowmelt, rainfall, and drought'; all familiar characters in stories of climate change.[31] Their methodology established a series of settings, including place, characters and actions. Utilising DiMaggio's definition of culturally available schemata as 'knowledge structures that represent objects or events',[32] they found that such schemata supported individuals to establish connections among narratives. This, in turn, enabled individuals to develop 'issue categorization frames, and cognitive schemas' that supported them to 'simultaneously hold multiple different narratives internally'.[33] They also demonstrated how providing a setting for a narrative strengthens its efficacy by ensuring a broader 'story world' that is based on the audience incorporating their own experience or knowledge.[34] Having a place where characters could interact operated as an effective mechanism for transporting the audience into the story in ways that could be more congruent with the audience's experiences and thereby more persuasive and memorable.

Having established the narrative setting, in this case, the flooding of communities along the Yellowstone River in Montana, Shanahan, Jones and McBeth coded sets of characters by the type of agency they held within a narrative, noting that nonhuman entities, such as the river, could also be a character, and referred to as having agency, emotions and personality. They employed three-character descriptors in their narrative analysis: 'A *hero* is the fixer of the problem. A *victim* is the entity being harmed. A *villain* is the entity causing the problem and/or inflicting harm'.[35] They also noted that 'The same entity can be a hero in one setting and a villain in another'.[36] They conducted semi-structured interviews with decision-makers from six municipalities within five of the ten counties through which the Yellowstone River flows. They explored responses through six interview questions framed by key concepts of experience and setting, the definition of the problem, who or what was the villain, the victim, or the hero, and finally what was the policy solution. These were:

> (EXPERIENCE/SETTING): Flooding on the Yellowstone has been occurring over many years. In thinking about [community], can you describe a memorable flood event?
>
> (PROBLEM DEFINITION): What happened? What were the consequences?
>
> (VILLAIN): Who is to blame for the [flood consequences]?
>
> (VICTIM): Who was affected by the [flood consequences]?
>
> (HERO): Who helped during this flood event? Who helped right after the event?
>
> (POLICY SOLUTION): Can you tell me about what plans were in place to prepare for that flood event?[37]

The interviews were coded first for narrative elements and then for issue categorisation frames. In a later study, Shanahan et al narrowed the focus to explore how the function of characters within a narrative relates to their value in risk communication.[38] Here, they found that for characters in action segments, narrative transportation which identifies the depth of engagement with and immersion in the story, reveals a largely negative total affective response for the victim narrative, a positive response for the hero narrative, and a more strongly positive response for the victim-to-hero narrative. Both the hero narrative and the victim-to-hero narrative have significantly higher total narrative transportation in this segment than does the victim narrative.[39]

They concluded that:

> When characters are first introduced and the problem of flooding is framed, we do not find audiences keenly engaged. However, when the drama of characters in action is presented, we see participants' affective responses intensify in distinct directions, depending on the characters depicted. Both narratives with heroes reveal increasing positive affective responses, with a sharper and higher positive affective increase for the victim-to-hero narrative; the victim narrative dips toward negative affective responses. Thus, the drama of characters in action matters in audience experience of narrative transportation.[40]

Shanahan et al also concluded that conventional risk communication is often 'ineffective at inducing people's preparedness' because 'Scientists tend to rely on the language of probability, uncertainty, frequency, and magnitude, whereas most people communicate their realities through stories, replete with characters, plotlines, and settings'.[41] This has implications in communicating the risk posed to cultural heritage by climate change, as well as the value of good conservation planning. Narrative framing can provide opportunities for individuals engaged in protecting a community's heritage to be cast as hero, breaking down the language of 'probability, uncertainty, frequency and magnitude' and replacing this with a language of agency, action and success. There is also scope for the local historical society or the photograph of the township prior to the bushfire, to take on victim-to-hero status.

A model for narrative enquiry

While some work has been done on mapping, assessing and understanding responses to climate change and the impacts of climate change on heritage places and objects,[42] the need for conservators and heritage professionals to engage with public policy debate, and to do so using evidence-based research, is pressing. At the same time, in the face of increasing large scale, catastrophic events stemming from climate change the main actors with significant agency are the individuals and communities for whom their heritage matters. For these reasons, the question becomes:

> What role does cultural conservation and heritage preservation play in supporting narratives that provide positive experiences?

And, following this:

> What are the research methods that are best suited to support this challenge?

Adaptation, resilience and mitigation have become three cornerstones of responses to the climate emergencies. They are explored in more detail in later chapters, however, all three rely on the effectiveness of narratives in which they are to be considered. In Small Island nations arguments that the need for relocation is a form of adaptation ring hollow against the reality that many rich nations refuse to adapt to a low-emission economy. For many communities, the scale of loss and change that results from the climate crisis is already affecting their ability to manage their heritage. Hearing and understanding the stories of individuals and communities who are already having to make decisions about the future of their heritage is a project of global significance. A keyword search of documents, such as those published by the various Intergovernmental Panel on Climate Change or the United Nations' Sustainable Development Goals or their related thematic issues, found virtually no overt mention and little implied emphasis on heritage. Yet for those working with communities affected by climate change, culture and heritage frame so much that is important in effective adaption, resilience and mitigation both before and after traumatic climate-crisis-induced events.

In Chapter 8, respondents from heritage institutions and organisations to a survey exploring responses to climate change, discussed how their organisations were publicly addressing the issue. The results indicated broad discrepancies. While significant research remains to be done on how heritage experts can have more impactful engagement within the global climate crisis response, telling, listening and collecting stories about heritage-related impacts of, and responses to climate change, is a useful extension of current community-engaged methodologies. Cultural materials conservation and other heritage-related disciplines are deeply interdisciplinary. As they continue to strengthen their focus on community engagement, they are increasingly drawing on social sciences to inform research and the translation of research into policy formulation and strategic action. Using narrative framework theory provides one tool from the social sciences that could be incorporated as a tool in heritage preservation. Adapting a model of the kind used by Shanahan et al has the potential to help heritage professionals devise compelling stories about heritage conservation and climate change.

One possible model could be based on a three-part approach that focuses on communication framed around the following three actions:

1. Be clear about what message you want to send. Identify what is your key message about heritage and climate change that will most engage stakeholders. Draft it up as a short, succinct statement.
2. Check what messages you are sending about climate change (this may be by omission of not including anything about climate change; there may be very specific messages you have made about climate change). Review your messages and check their alignment to your key message.
3. Examine how are you including examples of heritage heroes in climate change in your work. Check that your examples tell a compelling story and that you use this story effectively (Fig. 1.3).

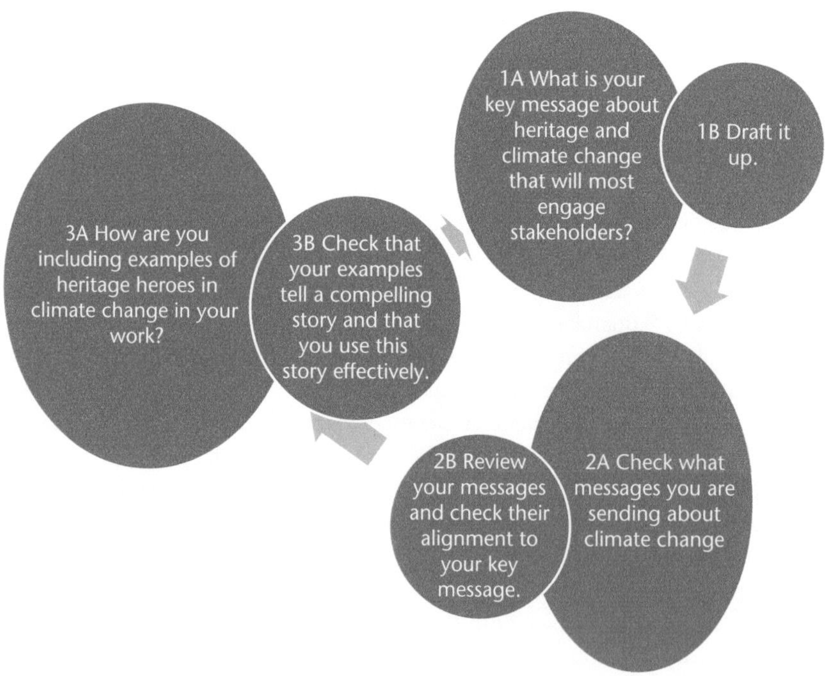

FIGURE 1.3 Flowchart of narrative enquiry for heritage preservation

Conservation and heritage preservation employ narrative construction in many formal processes including significance assessment, risk management, condition reporting and routine decision-making, but these tools have yet to be rigorously or consistently applied to questions about the impacts of the climate crisis on cultural heritage. It is partly for this reason that this book is framed by stories and storytelling. The other part of the reason for this is because this book has a focus on Australasia and the Indo-Pacific, and for many heritage experts in the region, stories hold heritage.

Incorporating a narrative framework into heritage conservation research enables research outcomes to be effectively communicated in ways that are powerful, credible and grounded. Translating this model into a heritage conservation framework is relatively straightforward. This framework can then be easily aligned to work on risk management, significance assessment, condition reporting, research, education or other disciplinary or professional frameworks.

This chapter explored how narratives have the power to shift public opinion, including people's feelings of being overwhelmed by the challenge of climate change impacts, and how refocussing narratives is an effective tool in climate change disaster preparedness and response. It argues that stories are particularly important in the management of heritage; first, because they are already embedded in professional methodologies such as significance assessments and conservation planning; second, because as a form of heritage their efficacy is well understood within the sector; and third, because, as Narrative Policy Framework theory demonstrates they are effective in engaging individuals and public opinion in critically important, and highly contentious discussions.

The following chapter applies this approach to various forms of heritage, including documentary heritage, oral histories, and local knowledge to show the value of co-ordinated, affective policy and strategy development to improve our understanding of, and ability to respond to climate change risks to cultural heritage.

Notes

1 Barnett, J, Tschakert, P, Head, L & Adger, WN 2016, 'Commentary: a science of loss', *Nature Climate Change*, Vol. 6, pp. 976–8.
2 Yeh, YC 2011, 'Ecological systems model of creativity development' in MA Runco & SR Pritzker (eds) *Encyclopedia of Creativity* (2nd ed.), Academic Press, Amsterdam, p. 293.
3 Apaydin, V 2020, 'The interlinkage of cultural memory, heritage and discourses of construction, transformation and destruction' in V Apaydin (ed) *Critical Perspectives on Cultural Memory and Heritage: Construction, Transformation and Destruction*, UCL Press, Los Angeles, pp. 13–29.
4 Harvey, D & Perry, J 2015, *The Future of Heritage as Climates Change: Loss, Adaptation and Creativity*, Taylor & Francis Group, United Kingdom, p. 10.
5 Harvey & Perry 2015, p. 7.
6 Holtorf C, 2018, 'Embracing change: how cultural resilience is increased through cultural heritage', *World Archaeology*, Vol. 50, No. 4, p. 648.
7 Barnett et al 2016, p. 977.
8 Barnett et al 2016, p. 977.

9 Carmichael, B, Wilson, G, Namarnyilk, I, Nadji, S, Brockwell, S, Webb, B, Hunter, F & Bird, D 2018, 'Local and Indigenous management of climate change risks to archaeological sites', *Mitigation and Adaptation Strategies for Global Change*, Vol. 23, No. 2, p. 2.
10 Cunsolo, A & Ellis, NR 2018, 'Ecological grief as a mental health response to climate change-related loss', *Nature Climate Change*, Vol. 8, p. 276.
11 See, for example, United Nations Office for Disaster Risk Reduction 2015; Carvallo 2014; Enenkel et al 2017; Sim et al 2017.
12 Sawitri, Hadiyanto & Hadi 2015 cited in Clayton, S, Manning CM, Krygsman, K & Speiser, M 2017, *Mental Health and our Changing Planet: Impacts, Implications, and Guidance*, American Psychological Association and ecoAmerica, Washington, DC, p. 6.
13 Clayton et al 2017, p. 17.
14 Clayton et al 2017, pp. 42–44.
15 Clayton et al 2017, p. 17.
16 Clayton et al 2017, p. 17.
17 See Russell, R & Winkworth, K 2009, *Significance 2.0: A Guide to Assessing the Significance of Collections*, Collections Council of Australia, Canberra ACT; Australia ICOMOS 2013a, *The Burra Charter: The Australia ICOMOS Charter for Places of Cultural Significance*, p. 10; Australia ICOMOS 2013b, *Preparing Studies and Reports: Contractual and Ethical Issues*, Burra Charter Practice Note, Version 1: November 2013, p. 9; and ICCROM & Canadian Conservation Institute 2016, *Guide to Risk Management*, https://www.iccrom.org/publication/guide-risk-management, p. 20.
18 The 1981 Australian Census recorded 768 people living in Cockatoo.
19 *ABC News*, 2009, 'Brumby warns of worse to come', 9 February 2009., accessed on 10 July 2022, https://www.abc.net.au/news/2009-02-09/brumby-warns-of-worse-to-come/289332
20 Teague, B, McLeod, R & Pascoe, S 2010, *Final Report Summary*, 2009 Victorian Bushfires Royal Commission, Parliament of Victoria, Melbourne, p. vii.
21 Teague, McLeod & Pascoe 2010, p. vii.
22 Teague, McLeod & Pascoe 2010, p. vii.
23 Werber, J & Lyons, S 2020, 'The size of Australia's bushfire crisis captured in five big numbers', *ABC Science*, 4 March 2020.
24 Thunberg, G 2019, 'Our house is on fire', presentation to the World Economic Forum, YouTube, 20 September 2019, Davos, https://www.youtube.com/watch?v=U72xkMz6Pxk
25 Cunsolo, A & Ellis, NR 2018, 'Ecological grief as a mental health response to climate change-related loss', *Nature Climate Change*, Vol. 8, p. 279.
26 Fisher, W 1984, 'Narration as a human communication paradigm: the case of public moral argument', *Communication Monographs,* Vol. 51, No. 1, pp. 1–22, https://doi.org/10.1080/03637758409390180
27 Fisher, W 1984, p. 16.
28 Patterson, M & Monroe, KR 1998, 'Narrative in political science', *Annual Review of Political Science*, Vol. 1, No. 1, p. 320.
29 Somers & Gibson 1998, cited in Patterson & Monroe 1998, pp. 325–6.
30 Patterson & Monroe 1998, p. 328.
31 Shanahan, E, Raile, E, French, K & McEvoy, J 2018, 'Bounded stories', *Policy Studies Journal*, Vol. 46, No. 4, p. 930.
32 DiMaggio, P 1997, 'Culture and cognition', *Annual Review of Sociology*, Vol. 23, p. 269 as cited in Shanahan et al 2018.
33 Shanahan et al 2018, p. 942.
34 Shanahan et al 2018, p. 930.
35 Shanahan, Jones and McBeth 2017, p. 335.
36 Shanahan et al 2018, p. 932.
37 Shanahan et al 2018, p. 948.

38 Shanahan, E, Reinhold, A, Raile, E, Poole, G, Ready, R, Izurieta, C, McEvoy, J, Bergmann, N, & King, H 2019, 'Characters matter: how narratives shape affective responses to risk communication', *PLoS One*, Vol. 14, No. 12, p. 1–24.
39 Shanahan et al 2019, pp. 18–19.
40 Shanahan et al 2019, p. 21.
41 Shanahan et al 2019, p. 4.
42 See Sandford & Cassar 2020; Pagliarino & Meredith 2020.

References

ABC News 2009, 'Brumby warns of worse to come', 9 February 2009, https://www.abc.net.au/news/2009-02-09/brumby-warns-of-worse-to-come/289332

Apaydin, V 2020, 'The interlinkage of cultural memory, heritage and discourses of construction, transformation and destruction' in V Apaydin (ed) *Critical Perspectives on Cultural Memory and Heritage: Construction, Transformation and Destruction*, UCL Press, Los Angeles, pp. 13–29.

Australia ICOMOS 2013a, *The Burra Charter: The Australia ICOMOS Charter for Places of Cultural Significance.*

Australia ICOMOS 2013b, *Preparing Studies and Reports: Contractual and Ethical Issues*, Burra Charter Practice Note, Version 1, November 2013.

Barnett, J, Tschakert, P, Head, L & Adger, WN 2016, 'Commentary: a science of loss', *Nature Climate Change*, Vol. 6, pp. 976–978.

Carmichael, B, Wilson, G, Namarnyilk, I, Nadji, S, Brockwell, S, Webb, B, Hunter, F & Bird, D 2018, 'Local and Indigenous management of climate change risks to archaeological sites', *Mitigation and Adaptation Strategies for Global Change*, Vol. 23, No. 2, pp. 231–255.

Cavallo, A 2014, 'Integrating disaster preparedness and resilience: a complex approach using System of Systems', *Australian Journal of Emergency Management*, Vol. 29, No. 3, pp. 46–51.

Clayton, S, Manning CM, Krygsman, K & Speiser, M 2017, *Mental Health and Our Changing Planet: Impacts, Implications, and Guidance*, American Psychological Association and ecoAmerica, Washington, DC.

Cunsolo, A & Ellis, NR 2018, 'Ecological grief as a mental health response to climate change-related loss', *Nature Climate Change*, Vol. 8, pp. 275–281. doi:10.1038/s41558-018-0092-2.

DiMaggio, P 1997, 'Culture and cognition', *Annual Review of Sociology*, Vol. 23, p. 263–287.

Enenkel, M, Papp, A, Veit, E & Voigt, S 2017, 'Top-down and bottom-up—A global approach to strengthen local disaster resilience,' *2017 IEEE Global Humanitarian Technology Conference (GHTC)*, 2017, pp. 1–7. doi:10.1109/GHTC.2017.8239247

Fisher, W 1984, 'Narration as a human communication paradigm: the case of public moral argument', *Communication Monographs*, Vol. 51, No. 1, pp. 1–22. doi:10.1080/03637758409390180

Harvey, D & Perry, J 2015, *The Future of Heritage as Climates Change: Loss, Adaptation and Creativity*, Taylor & Francis Group, United Kingdom.

Holtorf C, 2018, 'Embracing change: how cultural resilience is increased through cultural heritage', *World Archaeology*, Vol. 50, No. 4, pp. 639–650. doi:10.1080/00438243.2018.1510340

ICCROM & Canadian Conservation Institute 2016, *Guide to Risk Management*, p. 20, https://www.iccrom.org/publication/guide-risk-management

Pagliarino, A & Meredith, A 2020, 'Mapping climate change and risks for Australian cultural collections', *AICCM Bulletin*, Vol. 41, No. 1, pp. 3–26.

Patterson, M & Monroe, KR 1998, 'Narrative in political science', *Annual Review of Political Science*, Vol. 1, No. 1, pp. 315–331. doi:10.1146/annurev.polisci.1.1.315

Russell, R & Winkworth, K, 2009, *Significance 2.0: A Guide to Assessing the Significance of Collections*, Collections Council of Australia, Canberra ACT.

Sandford, R & Cassar, M 2020, 'Heritages of futures thinking: strategic foresight and critical futures' in C Holtorf & A Hogberg (eds) *Cultural Heritage and the Future*, Routledge, London, pp. 245–263.

Shanahan, E, Jones, M & McBeth, M 2017, 'How to conduct a Narrative Policy Framework study', *The Social Science Journal*, Vol. 55, No. 3, pp. 322–345. doi:10.1016/j.soscij.2017.12.002

Shanahan, E, Raile, E, French, K & McEvoy, J 2018, 'Bounded stories', *Policy Studies Journal*, Vol. 46, No. 4, pp. 922–948. doi:10.1111/psj.12269

Shanahan, E, Reinhold, A, Raile, E, Poole, G, Ready, R, Izurieta, C, McEvoy, J, Bergmann, N, & King, H 2019, 'Characters matter: how narratives shape affective responses to risk communication', *PLoS One*, Vol. 14, No. 12, pp. 1–24. doi:10.1371/journal.pone.0225968

Sim, T, Dominelli, L & Lau, J 2017, 'A pathway to initiate bottom-up community-based disaster risk reduction within a top-down system: the case of China', *International Journal of Safety and Security Engineering*, Vol. 7, No. 3, pp. 283–293.

Teague, B, McLeod, R & Pascoe, S 2010, *Final Report Summary*, 2009 Victorian Bushfires Royal Commission, Parliament of Victoria, Melbourne.

Thunberg, G 2019, 'Our house is on fire', presentation to the World Economic Forum, YouTube, 20 September 2019, Davos, https://www.youtube.com/watch?v=U72xkMz6Pxk

United Nations Office for Disaster Risk Reduction 2015, *Sendai Framework for Disaster Risk Reduction 2015–2030*, United Nations, https://www.un-spider.org/sites/default/files/43291_sendaiframeworkfordrren.pdf

Werber, J & Lyons, S 2020, 'The size of Australia's bushfire crisis captured in five big numbers', *ABC Science*, 4 March 2020.

Yeh, Y-C 2011, 'Ecological systems model of creativity development' in MA Runco & SR Pritzker (eds) *Encyclopedia of Creativity* (2nd ed.), Academic Press, Amsterdam, p. 293.

2

FINDING UNEXPECTED DATA FROM THE HISTORICAL RECORD

Marcelle Scott and Robyn Sloggett

This chapter explores how heritage sites and archival documents have been utilised in climate change research. It extends the idea of narrative as a method for activating agency and framing decision-making in cultural heritage conservation and preservation. In the volume *The Future of Heritage as Climate Change,*[1] David Harvey and Jim Perry conclude that the relationship between heritage and climate change is both processual and contingent ('heritage is a process and a product'), and dynamic (enabling 'a search for the creative possibility of transformation').[2] They argue that 'Heritage is also a critical issue for climate change. Perceptions of the past inform our understanding of climate change processes'.[3] This chapter explores this proposition through a number of examples. Extending the methodology introduced in Chapter 1, Chapter 2 considers a role for heritage in climate crisis research and response that is participatory and heroic. In this role, heritage 'products', as a core component of climate change studies, provide the critical data for climate change science and decision-making; and heritage methodologies provide the intellectual frameworks with which to understand how people value what they inherit from the past, protect this inheritance in the present, and aspire for their inheritance to pass into the future.

Archives, place and memory

The Sakura, the time when the cherry blossoms bloom in Japan, is a culturally significant event, celebrated with hanami, or viewing parties, and picnics. The Sakura has been documented in diaries and local chronicles stretching back some 1300 years. Sakura is a particular form of archive, integrating ancient records with contemporary practices and bringing the intangibility of expectation and

DOI: 10.4324/9781003163312-3

memory to the tangibility of place. It is a richly documented, lived heritage that demonstrates how an event triggers cultural behaviour. Such events are common around the world. In Bulgaria, for example, people celebrate Baba Marta Day recognising the first blooming of a tree for the new spring. In Ghana, the celebrations for Homowo, which roughly means 'to hoot at hunger', begin in May with the planting of millet and continue to August. In late September, the people of Impruneta in Italy, celebrate the grape harvest with La Festa dell'Uva.

In recent times, however, the dates for Sakura have been changing, and in 2021 Sakura in Kyoto reached full bloom on 26 March, the earliest date on record.[4]

In Tokyo (Edo), the Central Meteorological Observatory (now the Japan Meteorological Agency) had begun routine observations in 1876.[5] But it was much earlier records from Edo that enabled environmental scientist, Yasuyuki Aono, to explore phenological records for the 207-year period from 1601 to 1905 AD that provided the full blossoming dates for the cherry blossoms from *P. jamasakura*.[6] After studying these records, Aono concluded that '[b]ased on these data and a comparison with synchronous data from Kyoto, it was possible to obtain a reliable time series of temperature changes in Edo'.[7]

Celebrating Sakura is a lived heritage stretching back over a millennium. It is accompanied by rich collections of objects and archival documents that record these celebrations. Aono describes how scientists scrutinised some 150 historical records from the ninth century to chart the timing of cherry blossoms, and how these documents provided accounts of

> observations of cherry blossoms at the time of full flowering, sightseeing and parties held for the purpose of viewing the blossoming cherry trees, reports of gifts made from twigs containing cherry blossoms, and Japanese poetry composed about cherry blossoms.[8]

These records, preserved over generations by emperors, aristocrats, monks and others, enabled Aono, and colleagues Keiko Kazui and Shizuka Saito, to build a phenological profile of the timing of Sakura.[9] Their work was complemented by that of the Japan Meteorological Agency, which records data from 58 'bench-mark' flowering cherries from across Japan. As the timing of cherry blossom is heavily dependent on temperature these 58 trees provide important evidence for climate change studies. On the basis of this dataset Shunji Anbe, from the Agency's observation division, concluded that changes in flowering is ' … most likely because of the impact of the global warming'.[10] In this way, documentary heritage, natural heritage and intangible heritage have been brought together to provide a compelling argument that demonstrates the effects of climate change on the environment, and on things that people have held as immutable for millennia.

News of the early Sakura made headlines around the world in 2021. For local communities, the results have been disastrous. The celebration of Sakura brings tourists and income, but there is much more at stake. The shift of behaviour in one species is mirrored in similar impacts on others; and, as it is useful to be reminded, humans are only one of many species. The risk to the enjoyment of Sakura is only one of the impacts on the various forms of cultural expression that are associated with the Sakura, which has resulted from climate change. Climate change risks the cultural future of Sakura, as well as creating environmental changes that may risk the ability to properly care for the very cultural heritage that enables us to understand the magnitude of the shifts in climate that are threatening Sakura.

The fact that climate change data can be extracted from the distributed Sakura archives, from documents that have existed for over a millennium, speaks to the long tradition and embedded value of conservation research and practice, even before the Western concept of a conservator had been formulated. The Sakura archives are not unique. Public museums, archives, libraries, art galleries and private collections hold documents about climate, environment, weather and significant events, which have been preserved and remain available for use by researchers including artists, scientists, writers, local community historians and more. Neither are these documentary archival formats unique in containing valuable information for climate change studies. Buildings and sites that have been preserved over generations also hold data that help us understand climate change while at the same time standing as a legacy to what is valued.

The links between a place and the archives that relate to it are a feature of other climate change science projects. Port Arthur, as one of eleven Australian Convict Sites, is a World Heritage Site inscribed on the World Heritage List in 2010. Established in September 1830 as a woodcutting camp for convicts, by 1836 the settlement of Port Arthur had 1000 adult prisoners with another 300 juvenile convicts housed across the bay in the prison at Point Puer. By 1842, Port Arthur, with a flour mill and houses, was a significant centre for manufacturing. When the practice of transportation ceased in 1853 Port Arthur transitioned from a penal colony to a welfare establishment, providing accommodation for paupers, invalids and the mentally ill.[11] Through all this time records were made and kept that related to Port Arthur, its management and its environment. In a study of changing sea levels, Hunter, Coleman and Pugh examined records made at Port Arthur between 1841–1842, 1999–2002 and intermediate observations made in 1875–1905, 1888 and 1972.[12] The first record, made on 1 July 1841, consisted of a sea-level benchmark of one horizontal and three radiating vertical incised lines made on a cliff on the Island of the Dead immediately across the water from Port Arthur (Fig. 2.1).

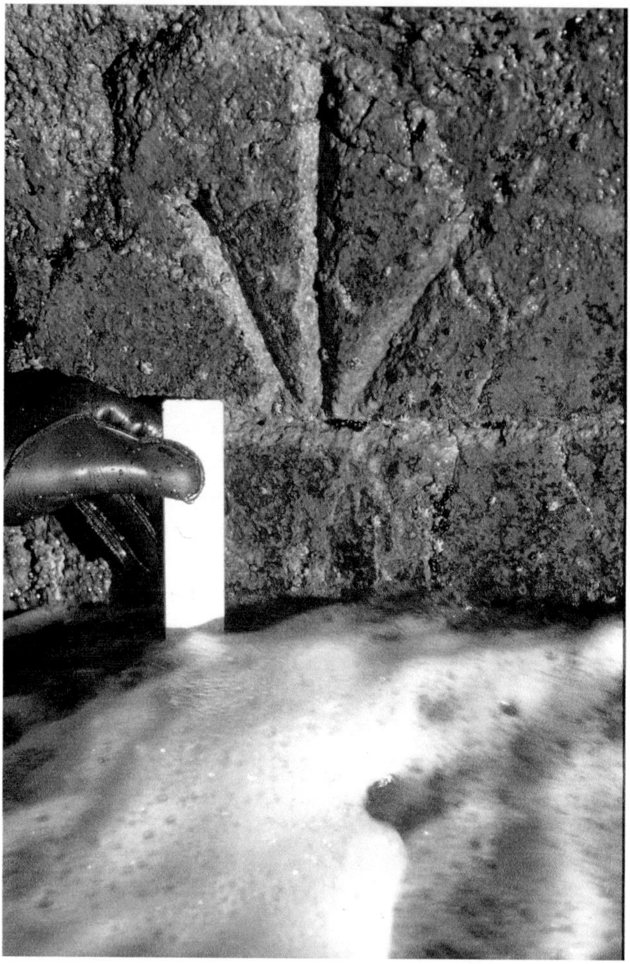

FIGURE 2.1 The sea-level benchmark carved in 1841 on the Isle of the Dead, Tasmania. Photo by Bruce Miller/Alamy

The second set of records is contained within diaries and records made by Thomas James Lempriere, who assiduously diarised the time he spent at Port Arthur. The Lempriere archives are now distributed across a number of organisations including the State Library of New South Wales that hold records from 23 March 1834 to 30 April 1836 and 1 January 1837 to 12 September 1838, the archives of the Royal Society in London which has records from 1841 and 1842, and the National Archives of Australia with records for December 1839 and February 1840–January 1841.[13] The third set of data came from the Aquatrak acoustic tide gauge, which was installed in the Port Arthur settlement in 1998. Exploring sea level rises in southeast Australia, Hunter, Coleman and Pugh estimated the sea level rise based on the difference between Lempriere's records for 1841–1842 and the Aquatrak acoustic tide gauge. They concluded that

> historic and modern records from Port Arthur, Tasmania, cover the longest time span of any sea level observations in the Southern Hemisphere and are related to a single benchmark; they provide a significant contribution to our knowledge of past sea level rise in this data-sparse region.[14]

Lempriere also collected natural history collections, specifically fish of which a number are preserved in glass bottles in the Natural History Museum, London.[15] Other institutions also hold material collected at Port Arthur. The Tasmanian Herbarium collection contains three botanical specimens that are recorded as having been collected at Port Arthur by naturalist Joseph D Hooker on 26 October 1840. Two specimens from Port Arthur and Eagle Hawk Neck collected by Hooker in the same month are held at the Royal Botanic Gardens, Kew. Another botanical specimen that was sent to Kew in 1833 had been collected at Port Arthur by the missionary James Backhouse.[16] Taken together these records provide information across a number of domains, including sea level and species distribution, all of which are of increasing interest to climate change scientists.

Although the data from Port Arthur is small, it is highly significant. Hunter, Coleman and Pugh note that estimates for the global average rate of rise in sea level, estimated to have been 1–2 mm/year over the twentieth century, are based on 24 tide gauges that are distributed unevenly across the globe. Of these, only six are in the Southern Hemisphere, and none are in Australia.[17] The data that is available from the dispersed historical records that originate from Port Arthur is therefore unique in being the only data available that records sea levels since European occupation of Australia.

The dispersed and often serendipitous nature of heritage data is acknowledged by Kari James, Executive Officer of the Pacific Manuscripts Bureau (PAMBU), who describes how the 4,000 reels of microfilm held by PAMBU representing 220,000 pages of digitised documents and photographs, are being used to provide information on climate change. The collection contains Church Mission Records, Whaling Records, Government Records, Diaries/Letters, Trading/Shipping Records, Dictionaries and Word Lists, Agricultural Records and Linguistic Records.[18] They range geographically and temporally. Examples include records from Eversleigh Station near Armidale in northern New South Wales in Australia from the period 1877–1922[19]; records made by Methodist missionary J E Moulton and students at Tupou College in Tonga from 1874 to 1878 (PMB Ms 985-10); the New Caledonian newspaper *Le Moniteur de la Louvelle Caledonie Noumea* from 1859 to 1886; and the Suwarrow Diary from a remote part of the Cook Islands that contains daily records of wind and barometric pressure for around three months from 1902. The PAMBU microfilm archives also contain images of landscapes that have altered over time, descriptions of past weather events, weather records, and a range of other descriptions and images, such as heatwaves, floods and droughts. These records provide the qualitative data that describes how societies were impacted and how they have responded. While programs such as PAMBU demonstrate the value of distributed archival records that are held in institutions and communities across the Australian, Asia-Pacific region, Kari James also identifies a number of issues associated with maintaining and using these particular datasets. Issues identified by James include that the metadata is not comprehensive; the cost of conversion from microfilm to digital impedes access; there is no cohesive interface between the international climate science community and international archives/library community; and to date few Indigenous researchers have engaged with the collection.[20] Similar issues are of concern to almost any archive and collection holding records that might be relevant to climate change studies, and the risk is that rare and significant information will be lost if these collections, often distributed, are not properly supported.

There are many other examples of locally based collections that hold information relevant to climate change studies. Near Norwich, in Norfolk in the United Kingdom, the archives of the Marsham family contain records held by the family for over 200 years, from 1736 to 1947. This sustained body of information contains evidence of how species respond to changes in climate in one specific location. These family archives identify 'first dates of observation, or "indications of spring" for 27 phenological events that relate to over 20 species of plants and animals'.[21] Similar records exist, albeit most on a lesser scale, in historic houses and family archives across the globe. In their 2007 paper, Quansheng Ge and fellow researchers published a paper that addressed the challenge of attempting to develop climate change reconstructions from incomplete historical data.[22] They drew on sets of records comprising some 32,351 volumes and including, amongst others, Wu-Xing-Zhi (the Chronology of the Five Elements) which chronicled disasters and abnormal events, such as a solar eclipse; Yu-Xue-Fen-Cun (the Depths of Rainfall infiltration and Snowfall) records from the Qing Dynasty, which covered 273 prefectures in 18 provinces from 1693 AD to 1911 AD; Song-Shi (the History of the Song Dynasty) providing records from 960–1279 AD; and the Qing-Yu-Lu (Records of Sunny or Rainy Days) weather record with detailed information on sky conditions, wind direction, type and duration of precipitation from 1723 AD to 1903 AD from four locations: Beijing, Nanjing, Suzhou and Hangzhou. These records were assessed by their 'objectivity', that is as records that can be compared against a range of other sources and their 'subjectivity', that is, those that cannot easily be subjected to quantitative analysis or comparison. As part of objective data, the research group also included older natural-derived data such as tree rings and ice cores that enabled them to extend the timelines that were being studied.

Indigenous cultural records and climate change

The records identified above reflect a mix of recorded and practiced heritage traditions. Ge et al relied on physical evidence, tree rings and ice cores to extend the scope of analysis. In Australia, knowledge about climate that stretches back past the last Ice Age has been collected and held in communities over millennia. In Southeast Asia and the Pacific, long term and highly detailed knowledge about climate is still actively managed within communities. The Australian Bureau of Meteorology lists the seasons as identified by 17 different Aboriginal language groups.[23] This demonstrates the nuanced and complex understanding of local environments developed by Australian Indigenous societies over millennia. In southwest Australia, the Nyoongar people identify six different seasons in the annual cycle. Birak occurs from December to January, Bunuru from February to March; Djeran from April to May, Makuru from June to July, Djilba from August to September and Kambarang from October to November. The Gariewerd people of southeastern Australia also have six distinct seasons that differ slightly from those of the Nyoongar: Kooyang (late summer, January to March), Gwangai moronn (Autumn, March to May), Chunnup (Winter, May to July), Larneuk (Pre-spring, July–August), Petyan (Spring, September to November) and Ballambar (Early Summer, November to January). In northwestern Australia in Miriwong country, Miriwoong people identify three seasons; Nyinggiyi-mageny (the wet season from December, January, February and March), Barndenyirriny (the hot season from September, October and November) and Warnka-mageny (the cold season from April through to August), which have eight sub-seasons. The Masigalgal people in the Torres Strait identify four seasons; the Kuki (December to April); Zei (November to January); Woerr/Sagerr (March to September); and Naigai (September to November).[24]

This knowledge of seasonal change is important when managing the land. For example, many Australian Indigenous communities have 'Caring for Country' programs that engage local Ranger groups in bushfire prevention using traditional Indigenous burning techniques. These techniques require detailed knowledge of wind, temperature and humidity as well as the behaviour of species in fire situations. Across the world, communities preserve knowledge through performance, stories and practices that enable them understand their local weather and time their activities according to the seasons. In turn, these seasons denote the ways that the world behaves, and people are dependent on this knowledge for their survival.

The examples described above give some sense of the complexity of both the forms of records that are available and the ways in which they are used to investigate climate change. They highlight the disciplinary, interdisciplinary and cross-cultural nature of such investigations. They are less revealing about the preservation methodologies that were brought to bear to enable these records to continue to exist. The fact is that without the agency of various heritage experts, the professionals, Senior Knowledge holders and custodians who work to retain heritage, these records would not be available as a resource by which to explore climate change and mitigation, adaptation and resilience. There are very few publications that discuss the role of heritage experts in protecting heritage for future research. Rather, this role is seen as procedural instead of innovative, heroic or even significant. Nevertheless, archives such as PAMBU do not collect, maintain and preserve themselves, and as Kari James notes, there is substantial effort require to ensure collections are comprehensively catalogued, available and accessible in suitable format and condition, and available to those who can best use them.

Adequate resourcing is necessary to support conservation and preservation at the scale required to preserve the knowledge needed in the fight against climate change. Without a line of sight from the research programs that use heritage in their quest to uncover new knowledge about the climate and climate change, to the programs that ensure this heritage is available for use, such resourcing is unlikely to be forthcoming. This omission is impactful.

Using cultural heritage to identify risk

In exploring the impact of integrated historical climate change research using archival records, climate scientist Rob Allan and fellow researchers describe the Atmospheric Circulation Reconstructions over the Earth (ACRE) program, which 'links some 100+ projects, institutions, organisations, and data rescue and climate applications activities around the globe'.[25] Commencing in 2008 and still active, ACRE has provided the project framework for the recovery of historical instrumental surface terrestrial and marine global weather observations from the last 200–250 years, to underpin 3D weather reconstructions otherwise known as reanalyses. ACRE brings together projects and regional chapters from all continents to share data across the globe. This provides an extraordinarily valuable resource for research into climate change.[26] As upper air observations obtained by satellite data are only available from 1957, and satellite data has only been available since 1979, the data that the ACRE projects collect is particularly important for historic reanalyses, one of many methods that ensures that historic data is available for longitudinal studies in contemporary climate change science.[27] Reanalysis can supply large amounts of information at each step providing, for example, data on trends in modern global monsoonal rainfall on land, or time-series graphs of global temperatures at 2 metres from 1870 to 2013.[28] By utilising a clearly defined forecast model and assimilation method that enables large amounts of historical records to be ingested, researchers are able to reanalyse specific observations made by particular networks, such as old shipping logs or early farming records. As applied mathematician and data analyses researcher Laura Slivinski identifies:

> Putting specific events in a climatic context is central to understanding the present and predicting the future. This applies especially to extreme weather events such as floods, droughts, and hurricanes. Observational studies require a large enough sample of events to be able to control for other factors, so a centennial history of weather is necessary for these high-impact, low-frequency events.[29]

Recently Slivinski was one of 40 authors who reported on the NOAA-CIRES-DOE 20th Century Reanalysis Version 3 (20CRv3), a dataset that provides global, three-hourly estimates of the atmosphere from 1806 to 2015 that 'can skillfully represent mass, circulation, and precipitation fields, on synoptic to climatic scales'.[30] The scale of this enterprise is breath-taking, yet it is the individual data sets, recorded, transcribed and uploaded that provide the capability for this work.

Other programs similarly draw down on the archival record. The EU Surface Temperature for All Corners of Earth (EUSTACE) is a HORIZON 2020 project involving eight institutions in four countries. It includes over 35,000 daily series of maximum and minimum temperatures obtained from seven different collections of station and satellite observations of which 97% are publicly available.[31] By April 2018, EUSTACE had created a data set of around 750 million unique observations over the period 1850–2015.[32] Philip Brohan, Climate Scientist at Met Office Hadley Centre in Exeter, Devon developed a video imaging of the EUSTACE data sets that provides a quick visualisation of daily temperatures taken from in situ observations.[33] As well as being important outcomes of the scientific research programs, visualisations such as these provide informative snapshots that are readily accessible and provide the public with a sense of the science behind climate change.

These highly complex research programs are all predicated on core data that exists in hard copy formats, on paper, as photographs or paintings, and the like. In an article in *Australian Geographic* from November 2020 Rob Allen rang alarm bells when he claimed 'It's hard to know for sure, but probably only half of the known historical weather observations have been digitised so far, and the world loses about 500,000 old records each day'.[34] Adding to Kari James' discussion of the problems in using archives for climate change research, Allan et al discuss additional challenges that researchers face including disciplinary separation that

> overlooks the cultural and social context of its production … [with] contrasting categories of research questions, different modes of academic debate, and different approaches to the presentation of research between disciplines.

They also point to the 'differences in ways of thinking and processing knowledge that underpin the language of, and research training in, different disciplines'.[35]

There are other research models for how heritage can be used to study climate change. Climate History Australia was a Citizen Science project that, by 2021, had involved thousands of volunteers transcribing traditional climate knowledge. In 2021, volunteers were kept busy with the project 'Reconstructing Perth's pre-industrial climate' that involved transcribing, collating and digitising records relating to weather events in Perth in Western Australia that originated between 1876 and 1897. The records they were working on consisted of handwritten weather records, early newspapers, photographs, artworks, correspondence and accounts from early settlers and colonial record keepers. This project was undertaken by 1,800 volunteers who worked on Zooniverse, a web-based citizen-science platform, and who managed to fill in gaps left outstanding from a previous project that had examined Perth's weather from 1830 to 1975.[36] They worked on archives that included journals from the Perth Survey Office from 1880 to 1884 and the Western Australian Botanic Gardens from 1885 to 1900, and now kept at the National Archives of Australia (NAA). Another citizen science project is addressing the next challenge of filling the gaps in daily observations between 1876 and 1879. A similar Climate History Australia project 'Reconstructing Adelaide's pre-industrial climate' engaged over 1,100 citizen scientists, also on Zooniverse, who recovered over 33,400 valuable historical weather observations which filled an eight-year data gap in records for Adelaide. These two record sets established through standard collection management strategies, but focused by the climate crisis, now provide some of the longest data record sets available in the Southern Hemisphere.[37]

Science is everywhere, but it needs information to produce evidence. In climate change studies, it is knowledge drawn from the historic record that is now providing some of the most useful longitudinal data in our understanding of climate change and management of natural disasters.

Working towards a model of climate crisis conservation

Archives, artworks, photographs, sea-level markings on cliffs, letters, diaries, logbooks, encoded oral histories, and much more form part of the historical record that is being enlisted in attempts to understand climate change. Nevertheless, across the globe, our ability to use these heritage resources to understand climate change is at risk. There are a number of reasons why this is the case.

1. Many collections remain uncatalogued. As we can't use what we don't know, their use for climate change science is compromised and their value continues to be unrealised. This can be addressed by an international program of small collection cataloguing that would go some way to ensuring that these valuable resources are identified and available for research.
2. As more cultural, historic and scientific heritage is produced, and as reasons for loss become more complex, many communities face increasing difficulty in preserving their cultural heritage. As a result, the increasing risks to cultural heritage across Australia and the Asia-Pacific region remain unaddressed and in some cases are escalating.[38] Information asymmetry, which is an outcome of this situation, is disrupting the capacity to realise the value of locally held knowledge and is eroding measures to protect these valuable resources. The need to have a clear, united voice across the heritage sectors that speaks to the need for a regional strategy for heritage conservation is the first important step in developing policies and strategies that are affective and able to be delivered. This can be addressed by professional and community peak bodies and senior knowledge holders coming together under an identified program to begin the process of lobbying for policies and strategies that umbrella existing local politics and the uneven distribution of resources.
3. Risks from degradation, damage, loss, disassociation and a range of other impacts precipitated by climate change are difficult for heritage professionals

> to address for a number of reasons. These include lack of funding, limited numbers of professionals in the areas that are required, lack of appropriate language skills, lack of appropriate cultural context and so on. Citizen science projects enable vast amounts of data to be captured through well-managed programs that enable one professional to engage hundreds of thousands of people. For this reason, and the others identifed above, an international citizen science program of small collection risk assessment and collection condition surveying would go some way to ensuring that these valuable resources are better able to be preserved.

Heritage records, in all their forms, are performing a heroic function in climate change science, yet the resources necessary to ensure their existence are not necessarily available. Just as climate change scientists are creating programs that predicate a global framework, the time is probably right for heritage experts have a commensurate significant contribution to make.

This chapter explored the interlinked functions of communities and professionals to secure and make available evidence for use by climate change scientists. It provided examples of how cultural heritage, across all its forms (tangible and intangible, place-based, as object, as document and archive, digital and analogue, and as oral history and performance) contributes to our understanding of climate change by providing data and securing evidence. Following from Chapter 1, this chapter also provides an indication of some of the ways in which collections are at risk, from degradation, damage, loss, disassociation and a range of other impacts precipitated by climate change. It examined the need to improve the ways collections can be managed to enable them to be available for climate change research, including the need to improve metadata capture and storage; the need to manage costs of conversion (and emulation); and the need to support and grow the engagement of Indigenous researchers with collections, and researchers with Indigenous heritage experts. A number of examples demonstrated the value of understanding the cultural and social context in which heritage records and climate change science research operate. Finally, this chapter pointed to the potential problem of disciplinary separation that leaves heritage experts grappling with the problem of preservation and access while scientists seek comprehensive data sets of historic records.

Notes

1 Harvey, D & Perry, J 2015, *The Future of Heritage as Climate Change*, Routledge, Oxon.
2 Harvey & Perry 2015, p. 275
3 Harvey & Perry 2015, p. 3
4 Yamaguchi, M 2021, 'Japan's famous cherry blossoms bloom early as climate warms', *AP News*, 31 March 2021.
5 Yamaguchi 2021.

6 Aono, Y 2015, 'Cherry blossom phenological data since the seventeenth century for Edo (Tokyo), Japan, and their application to estimation of March temperatures', *International Journal of Biometeorology*, Vol. 59, No. 4, pp. 427–34.
7 Aono 2015, p. 433.
8 Aono 2015, p. 428.
9 See Aono, Y & Kazui, K 2008, 'Phenological data series of cherry tree flowering in Kyoto, Japan, and its application to reconstruction of springtime temperatures since the 9th century', *International Journal of Climatology*, Vol. 28, No. 7, pp. 905–14; Aono, Y & Saito, S 2010, 'Clarifying springtime temperature reconstructions of the medieval period for gap-filling the cherry blossom phenological data series at Kyoto, Japan', *International Journal of Biometeorology*, Vol. 54, pp. 211–9.
10 Yamaguchi 2021.
11 Mason, R, Myers, A & de la Torre, M 2003, *Port Arthur Historic Site*, Getty Conservation Institute, Los Angeles, p. 6.
12 Hunter J, Coleman, R & Pugh, D 2003, 'The sea level at Port Arthur, Tasmania, from 1841 to the present', *Geophysical Research Letters*, Vol. 30, No. 7.
13 Hunter, Coleman & Pugh 2003.
14 Hunter, Coleman & Pugh 2003.
15 Tuffin, R & Vertigan, C 2020, 'Shoot, catalogue, eat: interacting with nature at a Tasmanian penal station' in Australia ICOMOS, *Proceedings of the Science Heritage Symposium: Under the Microscope – Exploring Science Heritage*, Tasmania, 12 November 2018, pp. 1–8.
16 Tuffin & Vertigan 2020.
17 Church, JA, Gregory, JM, Hybrechts, P, Kuhn, M, Lambeck, K, Nhuah, MT, Qin, D & Woodworth, PL 2001, 'Changes in sea level' in JT Houghton & Y Ding 2001, *Climate Change 2001: The Scientific Basis*, Cambridge University Press, Cambridge, pp. 639–93; and Douglas, BC 1997, 'Global sea rise: a redetermination', *Surveys in Geophysics*, Vol. 18, No. 2–3, pp. 279–92, as cited in Hunter, Coleman & Pugh 2003.
18 James, K 2021, 'Exploring historical weather records in the PAMBU collection' [YouTube video], PARADISEC, 23 February, https://www.youtube.com/watch?v=ICOrBoGh1WM
19 Bridgman, H, Ashcroft, L, Thornton, K, di Gravio, G & Oates, W 2019, 'Meteorological observations for Eversleigh Station, near Armidale, New South Wales, Australia: 1877–1922', *Geoscience Data Journal*, Vol. 6, No. 2, pp. 174–88.
20 James 2021.
21 Sparks, TH & Carey, PD 1995, 'The response of species to climate over two centuries: an analysis of the Marsham phenological record, 1736–1947', *Journal of Ecology*, Vol. 83, No. 2, p. 321.
22 Ge, Q, Zheng, J, Tian Y, Wu, W, Fanga, X & Wang, WC 2007, 'Coherence of climatic reconstruction from historical documents in China by different studies', *International Journal of Climatology*, Vol. 28, No. 8, pp. 1007–24.
23 Bureau of Meteorology (BOM) 2016, 'Indigenous weather knowledge', http://www.bom.gov.au/iwk/
24 The Australian Bureau of Meteorology established the Indigenous Weather Knowledge website in 2002 in collaboration with the Aboriginal and Torres Strait Islander Commission (ATSIC) and Monash University's Centre for Indigenous Studies. 'The website is a formal recognition of traditional weather and climate knowledge that has been developed and passed down through countless generations by Aboriginal and Torres Strait Islander people. … Through the Indigenous Weather Knowledge website, the Bureau is working with communities that wish to record and share valuable seasonal and environmental information and traditional knowledge' (BOM 2016).
25 Atmospheric Circulation Reconstructions over the Earth (ACRE) 2020, 'What is ACRE?', http://www.met-acre.org/Home; Allan, R, Endfield, GH, Damodaran, V, Adamson, G, Hannaford, MJ, Carroll, F … Bliuc, A 2016, 'Toward integrated historical climate research: The example of Atmospheric Circulation Reconstructions over

the Earth', *Wiley Interdisciplinary Reviews: Climate Change*, Vol. 7, No. 2, pp. 164–74, https://doi.org/doi:10.1002/wcc.379
26 ACRE 2021, 'Data projects and regional chapters', http://www.met-acre.org/data-projects-and-regional-chapters.
27 Slivinski, LC 2018, 'Historical Reanalysis: What, How, and Why?', *Journal of Advances in Modeling Earth Systems*, Vol. 10, No. 8, pp. 1736–9.
28 Dee, D, Fasullo, J, Shea, D & Walsh, J 2016, 'The climate data guide: atmospheric reanalysis: overview & comparison tables', *National Centre for Atmospheric Research*, https://climatedataguide.ucar.edu/climate-data/atmospheric-reanalysis-overview-comparison-tables
29 Slivinski 2018, p. 1736.
30 Slivinski, LC, Compo, GP, Sardeshmukh, PD, Whitaker, JS, McColl, C, Allan, … Wyszyński, P 2021, 'An evaluation of the performance of the Twentieth Century reanalysis Version 3', *Journal of Climate*, Vol. 34, No. 4, p. 1431.
31 Brugnara, Y, Good, E, Squintu, AA, van der Schrier, G & Brönnimann, S 2019, 'The EUSTACE global land station daily air temperature dataset', *Geoscience Data Journal*, Vol. 6, No. 2, pp. 189–204.
32 Brugnara, Y, Auchmann, R, Brönnimann, S & Good, E 2018, 'The EUSTACE daily LSAT dataset: a collection of 35000+ unique in-situ daily temperature series with inhomogeneities identified', *Geophysical Research Abstracts*, Vol. 20, p. 7703.
33 Brohan, P 2020, 'EUSTACE and observations coverage', Vimeo, https://vimeo.com/403663259
34 Howlett, C 2020, 'Unearthing Australia's climate history', *Australian Geographic*, 11 November.
35 Allan, R, Endfield, GH, Damodaran, V, Adamson, G, Hannaford, MJ, Carroll, F … Bliuc, A 2016, 'Toward integrated historical climate research: the example of atmospheric circulation reconstructions over the earth', *Wiley Interdisciplinary Reviews: Climate Change*, Vol. 7, No. 2, p. 165.
36 Gergis, J, Baillie, Z, Ingallina, S, Ashcroft, L & Ellwood, T 2021, 'A historical climate dataset for southwestern Australia, 1830–1875', *International Journal of Climatology*, Vol. 41, pp. 4898–919.
37 Gergis, J n.d., 'Unearthing Australia's climate history', *Climate History Australia*, https://climatehistory.com.au/
38 See, for example, Markham & Richards 1933; Werner 1970; Pigott 1975; University Museums Review Committee 1996 and 1998; Deakin University 2002.

References

Allan, R, Endfield, GH, Damodaran, V, Adamson, G, Hannaford, MJ, Carroll, F, Macdonald, N, Groom, N, Jones, JM, Williamson, F, Hendy, EJ, Holper, PN, Arroyo-Mora, JP, Hughes, L, Bickers, RA & Bliuc, A 2016, 'Toward integrated historical climate research: the example of atmospheric circulation reconstructions over the earth', *Wiley Interdisciplinary Reviews: Climate Change*, Vol. 7, No. 2, pp. 164–174, doi:10.1002/wcc.379

ACRE 2021, 'Data projects and regional chapters', http://www.met-acre.org/data-projects-and-regional-chapters.

Aono, Y 2015, 'Cherry blossom phenological data since the seventeenth century for Edo (Tokyo), Japan, and their application to estimation of March temperatures', *International Journal of Biometeorology*, Vol. 59, No. 4, pp. 427–434, doi:10.1007/s00484-014-0854-0

Aono, Y & Kazui, K 2008, 'Phenological data series of cherry tree flowering in Kyoto, Japan, and its application to reconstruction of springtime temperatures since the 9th century', *International Journal of Climatology*, Vol. 28, No. 7, pp. 905–914.

Aono, Y & Saito, S 2010, 'Clarifying springtime temperature reconstructions of the medieval period for gap-filling the cherry blossom phenological data series at Kyoto, Japan', *International Journal of Biometeorology*, Vol. 54, pp. 211–219.

Atmospheric Circulation Reconstructions over the Earth (ACRE) 2020, 'What is ACRE?', http://www.met-acre.org/Home

Bridgman, H, Ashcroft, L, Thornton, K, di Gravio, G & Oates, W 2019, 'Meteorological observations for Eversleigh Station, near Armidale, New South Wales, Australia: 1877-1922', *Geoscience Data Journal*, Vol. 6, No. 2, pp. 174–188, doi:10.1002/gdj3.80

Brohan, P 2020, 'EUSTACE and observations coverage', Vimeo, https://vimeo.com/403663259

Brugnara, Y, Auchmann, R, Brönnimann, S & Good, E 2018, 'The EUSTACE daily LSAT dataset: a collection of 35000+ unique in-situ daily temperature series with inhomogeneities identified', *Geophysical Research Abstracts*, Vol. 20, https://meetingorganizer.copernicus.org/EGU2018/EGU2018–7703.pdf

Brugnara, Y, Good, E, Squintu, AA, van der Schrier, G & Brönnimann, S 2019, 'The EUSTACE global land station daily air temperature dataset', *Geoscience Data Journal*, Vol. 6, No. 2, pp. 189–204, doi:10.1002/gdj3.81

Bureau of Meteorology (BOM) 2016, 'Indigenous weather knowledge', http://www.bom.gov.au/iwk/

Church, JA, Gregory, JM, Hybrechts, P, Kuhn, M, Lambeck, K, Nhuah, MT, Qin, D & Woodworth, PL 2001, 'Changes in sea level' in JT Houghton & Y Ding (eds) *Climate Change 2001: The Scientific Basis*, Cambridge University Press, Cambridge, pp. 639–693.

Deakin University 2002, *A Study into the Key Needs of Collecting Institutions in the Heritage Sector*, Deakin University, Burwood.

Dee, D, Fasullo, J, Shea, D & Walsh, J 2016, 'The climate data guide: atmospheric re-analysis: overview & Comparison tables', *National Centre for Atmospheric Research*, https://climatedataguide.ucar.edu/climate-data/atmospheric-reanalysis-overview-comparison-tables

Douglas, BC 1997, 'Global sea rise: a redetermination', *Surveys in Geophysics*, Vol. 18, No. 2–3, pp. 279–292.

Ge, Q, Zheng, J, Tian Y, Wu, W, Fanga, X & Wang, WC 2008, 'Coherence of climatic reconstruction from historical documents in China by different studies', *International Journal of Climatology*, Vol. 28, No. 8, pp. 1007–1024, doi:10.1002/joc.1552

Gergis, J n.d., 'Unearthing Australia's climate history', *Climate History Australia*, https://climatehistory.com.au/

Gergis, J, Baillie, Z, Ingallina, S, Ashcroft, L & Ellwood, T 2021, 'A historical climate dataset for southwestern Australia, 1830–1875', *International Journal of Climatology*, Vol. 41, pp. 4898–4919, doi:10.1002/joc.7105

Harvey, D & Perry, J 2015, *The Future of Heritage as Climate Change*, Routledge, Oxon.

Howlett, C 2020, 'Unearthing Australia's climate history', *Australian Geographic*, 11 November, https://www.australiangeographic.com.au/topics/science-environment/2020/11/unearthing-australias-climate-history/

Hunter J, Coleman, R & Pugh, D 2003, 'The sea level at Port Arthur, Tasmania, from 1841 to the present', *Geophysical Research Letters*, Vol. 30, No. 7, doi:10.1029/2002GL016813

James, K 2021, 'Exploring historical weather records in the PAMBU collection', [YouTube video], PARADISEC, 23 February, https://www.youtube.com/watch?v=ICOrBoGh1WM

Markham, S & Richards, HC 1933, *A Report on the Museums and Art Galleries of Australia and New Zealand*, Museums Association, London.

Mason, R, Myers, A & de la Torre, M 2003, *Port Arthur Historic Site*, Getty Conservation Institute, Los Angeles. https://www.getty.edu/conservation/publications_resources/pdf_publications/pdf/port_arthur.pdf

Pigott, P 1975, *Museums in Australia*, Australian Government Publishing Service, Canberra.

Slivinski, LC 2018, 'Historical reanalysis: what, how, and why?', *Journal of Advances in Modeling Earth Systems*, Vol. 10, No. 8, pp. 1736–1739, doi:10.1029/2018MS001434.

Slivinski, LC, Compo, G P, Sardeshmukh, PD, Whitaker, JS, McColl, C, Allan, RJ, Brohan, P, Yin, X, Smith, CA, Spencer, LJ, Vose, RS, Rohrer, M, Conroy, RP, Schuster, DC, Kennedy, JJ, Ashcroft, L, Brönnimann, S, Brunet, M, Camuffo, D, Cornes, R, Cram, TA, Domínguez-Castro, F, Freeman, JE, Gergis, J, Hawkins, E, Jones, PD, Kubota, H, Lee, TC, Lorrey, AM, Luterbacher, J, Mock, CJ, Przybylak, RK, Pudmenzky, C, Slonosky, VC, Tinz, B, Trewin, B, Wang, XL, Wilkinson, C, Wood, K & Wyszyński, P 2021, 'An evaluation of the performance of the Twentieth Century reanalysis Version 3', *Journal of Climate*, Vol. 34, No. 4, pp. 1417–1438, doi:10.1175/JCLI-D-20-0505.1

Sparks, TH & Carey, PD 1995, 'The response of species to climate over two centuries: an analysis of the Marsham phenological record, 1736-1947', *Journal of Ecology*, Vol. 83, No. 2, pp. 321–329, doi:10.2307/2261570

Tuffin, R & Vertigan, C 2020, 'Shoot, catalogue, eat: interacting with nature at a Tasmanian penal station' in Australia ICOMOS, *Proceedings of the Science Heritage Symposium: Under the Microscope – Exploring Science Heritage*, Tasmania, 12 November 2018, pp. 1–8.

University Museums Review Committee 1996, *Cinderella Collections: University Museums and Collections in Australia*, Australian Vice Chancellor's Committee, Canberra.

University Museums Review Committee 1998, *Transforming Cinderella Collections: the Management and Conservation of Australian University Museums, Collections and Herbaria*, Australian Vice Chancellor's Committee, Canberra.

Werner, AE 1970, *Conservation of Cultural Property in Australia and the Territory of Papua New Guinea*, UNESCO, Paris.

Yamaguchi, M 2021, 'Japan's famous cherry blossoms bloom early as climate warms', *AP News*, 31 March, https://apnews.com/article/japan-cherry-blossoms-early-bloom-da90620656eee57d04f90d104640b261

3

MY COUNTRY IS CHANGING: INDIGENOUS PERSPECTIVES ON CLIMATE CHANGE

Robyn Sloggett, Marcelle Scott, and Gabriel Nodea

Increasingly, the unique knowledge that Indigenous communities and in particular Indigenous Elders hold is being recognised as having a valuable contribution to make to knowledge about climate change. This is recognised by UNESCO through its LINKS (Local Indigenous Knowledge Systems) programme which, in turn, aligns to the International Year of Indigenous Languages in 2019 following the adoption of the UN's 'Rights of indigenous peoples: resolution A/RES/71/178'.[1]

Across Southeast Asia and Australia, the continuous link between ancient knowledge systems and contemporary cultural practice is evidenced in the records of human occupation that date back millennia. In Sulawesi, in Indonesia, lithic scatters suggest occupation from at least 118,000 years,[2] with rock art dating back 45,000 years.[3] In Australia, similar evidence for human occupation has also been confirmed with dates reaching back to 65,000 years,[4] and rock art dating from 17,300 years ago.[5] It is not surprising that these earliest forms of representational art depict some of the animal species that were familiar to the the artists; the Warty Pig in Sulawesi and the kangaroo in Australia. These records are also evidence of continuous links between living cultural practice, place and the environment. While these rock art sites are tangible records of occupation and the knowledge associated with the place, most Indigenous knowledge is held in highly complex ontological encyclopedias that are held, maintained, shared and developed through language, performance, narrative and the association to land and landscape. This chapter examines some of the specific ways in which climate change is affecting the maintenance of these knowledge systems and how Indigenous knowledge is incorporated into Western-derived knowledge about climate change. It extends the scope of discussion in the previous chapter, which examined the types of records of the kind that UNESCO refers to as 'tangible', to explore Indigenous knowledge as a more integrated system of knowledge that complements Western scientific knowledge about climate change. In doing so, this chapter also locates

DOI: 10.4324/9781003163312-4

the concerns that many Indigenous communities have expressed about climate change and its effect on their lives and futures.

The inclusion of Indigenous knowledge in science education and research has increased markedly over the past decade and climate change studies are no exception. The UNESCO LINKS program comprises five themes: Science Policy, Biodiversity, Climate Change and Knowledge Transmission. The Knowledge Transmission project 'Revitalizing Knowledge – Indigenous Education' identifies 'an urgent need to enhance the intergenerational transmission of indigenous knowledge, as a complement to mainstream education'.[6] Science educators such as Handayani et al acknowledge the value of Indigenous knowledge in the curricula as 'an essential element of contemporary science knowledge'.[7] In Australia, the Bureau of Meteorology now includes Indigenous definitions of seasons and seasonal characteristics on its website.[8]

Leanne Cullen-Unsworth and Kirsten Maclean explore the alignment of Indigenous knowledge, place and evidence of climate change through the lens of cultural landscapes, arguing that the term 'cultural landscapes' defines both 'a reality, and a management approach' that demonstrates 'a more holistic view of the environment'.[9] This term goes some way to explaining the links between Indigenous ontologies, cultural practices and the immense care for country demonstrated by Indigenous communities, but such links are much more integrated than the qualified description as 'landscape' allows. Such knowledge is comprised of deep, scientific observational and experiential knowledge of country, species, weather and human behaviour, which is located and managed within a highly sophisticated and complex ontology. In a paper that examines 'Australian Aboriginal Ethnometeorology and Seasonal Calendars', Dr Philip Clarke argues that 'While Western style history serves to highlight the process of change, in Aboriginal worldviews it is the continuity and permanence'.[10] He cites anthropologist WE Stanner who described this interest that 'lies in the cycles rather than the continuum [of time]' as 'everywhen',[11] and contrasts this with 'Western European thought, [where] change and time is dealt with as both unidirectional and in cycles'.[12] Clarke also points to the difficulty European scholars have had in conducting ethnometeorological studies due to both the seasonality of weather observations and the difficulties in working across different cultural explanations and observations of weather. This in turn points to the limitations of Western research capability, and the need to develop new ways of thinking about Indigenous knowledge and knowledge preservation.

The Australian Bureau of Meteorology's Indigenous Weather Knowledge website provides an overview of the links between language, culture and environmental knowledge from a number of Indigenous communities across the country.[13] The website, which was established in line with the organisation's RAP (Reconciliation Action Plan), details the various seasons as identified by 17 different Aboriginal communities.

In southwest Australia, the Nyoongar people recognise six different seasonal cycles. Birak occurs over December and January. It is referred to as the season of the young and is characterised as dry and hot. It is the time when control burns are carried out in mosaics across the country to reduce fuel and stimulate food production. Bunuru, the season of adolescence, occurs over February and March and is the hottest and driest time of year. During this period, people inhabit the coast and freshwater plants and fish and seafood make up the major part of their diet. Djeran, in April and May, the season of adulthood, is when the dry breaks as the night air cools causing dew to collect on the ground and on plants. Like the other season's animals and plants also respond to the seasonal changes; the she-oaks develop heavily nectar-laden seed cones that provide food for mammals, birds and people and shelters are repaired and made waterproof in readiness for the increasingly wet weather. Makuru, occurring in June and July, is the season of fertility. It is the coldest and wettest time of the year when Nyoongar people leave the coast and move inland. Food becomes plentiful as the waterways begin to fill and animals and birds prepare for the breeding season. Djilba is the season of conception. It occurs in August and September when the weather is variable, with cold dry days and warm wet windy days interspersed with days of sunshine and good weather. Grazing birds and animals such as kangaroo, emus and possums are plentiful, and their young emerge at this time. Towards the end of the season temperatures start to climb. Kambarang, the season of birth, is the final season, from October and November. It is a transformational season when the country is rich with birds and animals, as they reproduce and care for their young. The weather is drying and temperatures are increasing. At the end of Kambarang, Birak commences and the cycle continues.

The Gariwerd people of southeastern Australia also have six distinct seasons that differ slightly to those of the Nyoongar: Kooyang (late summer, January to March), Gwangai moronn (Autumn, March to May), Chunnup (Winter, May to July), Larneuk (Pre-spring, July to August), Petyan (Spring, September to November) and Ballambar (Early Summer, November to January). These are characterised by the relationship between plants, animals, water, temperature, humidity and the sky. For example, Kooyang, the eel season, is the hottest and driest season when bushfires threaten the landscape and surface water is scarce. At night the stars are particularly bright. When it passes to the next season Gwangal moronn, the honey bee season, the days are warm but the mornings are cool and the sun rises red in the morning and sets with a golden glow in the evenings.

By comparison, in Miriwoong country in the eastern Kimberley region of northwestern Australia, there are three seasons – Nyinggiyi-mageny (the wet season from December, January, February and March), Barndenyirriny (the hot season from September, October and November) and Warnka-mageny (the cold season from April through to August) – divided into eight distinct sub-seasons. The Masigalgal people in the Torres Strait identify four seasons: the Kuki (December to April) which is storm time; Zei (November to January) a windy season when the wind rises from the southwest in the afternoon; Woerr/Sagerr (March to September) when a south easterly wind predominates; and Naigai (September to November) the hottest time of the year with calm steady winds.[14]

Such detailed knowledge is held in communities across the globe. Ingesting this as information in climate crisis studies and crisis management is a much more complex undertaking than the programs of data ingestion described in the previous chapter. Nevertheless, at the local level seasons denote the ways that the world behaves, and people are dependent on this knowledge for their survival.

On Gija land

The 17,300-year-old kangaroo that is evidence for the oldest dated rock art in Australia lived in the Kimberley region of northwest Australia.[15] The Kimberley is a large region of around 423,517 km^2. It is diverse, stretching north from the coastline along the Timor Sea, down into the Tanami Desert and Great Sandy Desert some 600 km to the south. From the west, the Kimberley stretches from the coast of the Indian Ocean, across some 1200 km to the east. The area was formed 1,800 million years ago when separate land masses collided to form the continent of Australia. Fossilised coral and Gogo fish fossils can be found as evidence of a vast coral reef that existed some 400 million years ago, when sea levels were much higher.[16] The effects of the end of the last Ice Age are also evident in the Kimberley. This was the time, around 11,700 years ago when the Pleistocene Epoch, which lasted from around 2.6 million to 11,700 years ago, came to an end. During the Pleistocene, sea levels dropped by up to 120 metres below present levels. According to Western science, the later part of the Pleistocene was the time when humans moved from Asia to Australia. A date of around 50,000 years has been proposed for continuous human occupation of the Kimberley.[17] This means that the Indigenous people of the Kimberley region occupied the land during the last Ice Age.

Today the Kimberley is characterised as a dry tropical zone with rain occurring from November to April, a period known as the wet season and characterised by heat and humidity, with heavy rain and intermittent cyclonic conditions that result from tropical depressions, thunderstorms and monsoons. Often floods will isolate communities and make the country impassable for weeks at a time. By contrast, the dry season which occurs from May to October is characterised by stable weather patterns, with clear intensely blue skies and cool nights. Early in the dry average maximum daytime temperatures are around 31°C and nightly minimums are around 18°C. The country is still green after the wet. Over the next two months temperatures drop about 2° to 3°C. As the dry progresses, rainfall decreases, the country dries up, and the colours of the iron-rich red and yellow ochres predominate. By October temperatures increase to an average of 37° to 38°C maximum during the day, dropping to lows of around 23°C at night. They continue to climb as the heat and humidity builds up in readiness for the wet season. As the wet builds up temperature and humidity increases, and storms and torrential rain fill the waterways. This environmental pattern of wet and dry is associated with very specific behaviours in the flora and fauna of the region, which demonstrate millennia of evolution and adaption.

The Gija people who live in the East Kimberley and whose land stretches some 32,000 km^2 have a history that extends back into what Western scientists call the

Pleistocene Epoch, but which, for the Gija people, is the Ngarranggarni, the time when the Ancestors travelled across the land giving name and form to everything. Ngarranggarni is the knowledge system that guides the Gija way of life. It was established by Gija Ancestors as they created the land, law, customs, plants, animals and people. The memory and knowledge of the Ancestors is held as a deep and rich ontology that describes the relationship between the land and everything in it. The Ngarranggarni encodes Gija belief and knowledge and proscribes the laws for living as a Gija person. The Ngarranggarni encompasses a broad and deep understanding of the past, present and future. It exists across all three simultaneously. The Ngarranggarni prescribes kinship relationship and accountabilities that exist between people, animals and plants. It guides contemporary Gija life by proscribing the proper way to conduct family and clan relationships and relationships with animals, plants and places in the ways that were established by the Ancestors when they formed the country. As a lived belief system, it ensures that Gija people understand their connection to Gija country, and their direct responsibility for their clan country.[18] Plants and animals have held their 'skin' names since the creation time when they took their present form. Skin names are given to people at birth. Within the Ngarranggarni Gija people are born into one of eight 'skin' groups, and within one of two cycles (moieties), each comprising four mother skin names and four father skin names.[19] All Gija people belong to one of these groups. Families, as traditional owners of country, manage protection and preservation of country and culture through inheritance. Each descent group owns, and is responsible, for a specific part of Gija country, and for maintaining the culture and for caring for the plants, animals and land associated with that country. In this way, responsibility for culture, land and people is practiced, taught and passed on to new generations through ceremony and cultural practice, and is integrated into every aspect of Gija community life. Places, objects, words, behaviours, songs, dance, plants, animals and people all work in a complete ontology that is designed to keep country strong and the world safe.

Gija Elders, who are respectfully referred to as the Old People, are responsible for ensuring the correct knowledge in regard to Ngarranggarni, and all the behaviours that flow from respect for Ngarranggarni knowledge. In particular, they are responsible for teaching younger generations. This knowledge cannot be passed on without the permission of the Old People.

In the 1970s Gija people, who were forced from their country by cattle station owners, created the township of Warmun, which today is an Aboriginal community of around 400 people situated on Gija lands in the East Kimberley. At that time the Old People established a school, in partnership with the Catholic order of teaching nuns, the St Josephite Order.[20] Known as the Ngalangangpum (Mother and Child) school, the Old People shared the formal school curriculum with its Western curriculum so that they could teach young Gija students about their country, law, customs, skin relationships, animals, plants and their responsibilities to look after the Gija world. The school children were also taught how to use plants and animals for food and to make tools and ceremonial items. Part of this teaching involved the creation of artwork, which was used to explain the Ngarranggarni stories associated with each Ngarrangganri place to the children who could not travel to their clan country. These paintings were also used to teach them about their responsibilities for caring for country. In 1998, the Old People established the Warmun Art Centre to recognise and respond 'to the need for a community owned and controlled centre through which they could support, maintain and promote Gija art, language and culture'.[21]

Climate change in context

Shirley Purdie, a senior Gija artist, Elder and co-author of the book *Jadagen Warnkan Barnden: Changing Climate in Gija Country,*[22] had a solo exhibition at the National Portrait Gallery, Canberra in 2020, which included the painting titled Goonjal (Fig. 3.1). She describes her painting for the online exhibitin catalogue:

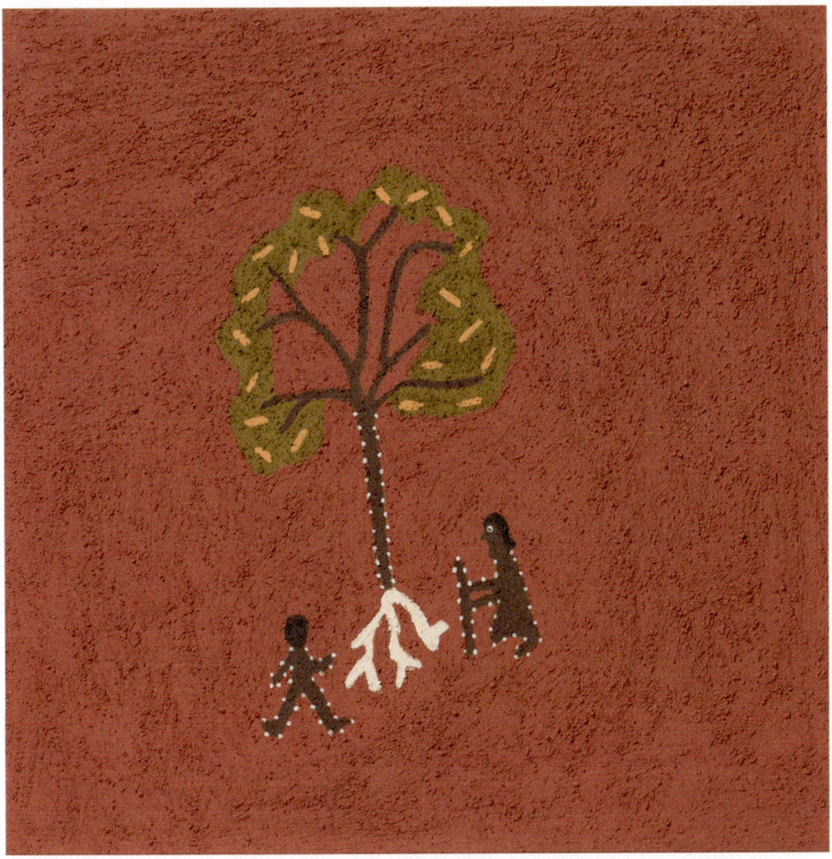

FIGURE 3.1 Shirley Purdie Goonjal (bush yam tree), 2018, 45 x 45 cm © The Artist. This is a picture of me and my mum digging yam in the time to dig yam, which is the cold weather. The yellow flowers remind us of the time for digging and eating yam

> This is that bush yam. This is my mother (right) and he bin teach me (left) for diggin' when I bin a little girl. This tucker now, this potato or root. When this tree gets yellow, that's the time we go and dig him.[23]

The online catalogue included further information about the painting:

> This is bush tucker knowledge Shirley's mother passed down to her. When she was a young girl, around ten years old, her mother would teach her how to dig up the root of the bush yam tree. She learned that when the tree flowered yellow, it was the correct time to dig up the roots. Generally, this happens in 'wangan' or cold weather/winter. However, Shirley has noted that climate change is affecting the seasonal flowering, and sometimes this tree bears yellow flowers when it is not supposed to.[24]

Linguist Frances Kofod, who has worked with the Gija Old People for over 40 years, observed that 'Detailed ecological knowledge is enriched by its placement in a cultural context'.[25] In Warmun, the context is changing quickly. Gija Cultural Officer, and past Chairman of the Warmun Art Centre, Gabriel Nodea said: 'It worries them [the Old People] a lot, it has for a few years'.

He continued:

> For some decades they know things are different. The kinds of critical predictions that can be made by knowing what is happening on the land, and by understanding the changes that have observed, are now much hard to make.
>
> The weather changes, animal behaviour and the landscape and what is happening on the ground; flower and trees being change. Critical prediction became hard [it was difficult] to tell what is going on the land. … They couldn't tell the difference, if it was bream season, or kangaroo season, or turkey season, or cold weather, hot weather, autumn, spring. And the damage was natural, caused by weather and disaster, and Mother Nature.
>
> Now because of global warming it's very hard to predict what is going on.

The escalating impact, according to Nodea, was that people 'didn't want to go out anymore'. It became hard to 'tell exactly what Mother Nature has to provide for us'. This inability to predict what the climate is doing impacts on people's confidence in traditional life:

> We don't know what's going on with climate change. We know that it's like some sort of another big difference that come across the land and destroys the land, and make it hard for us to live our … days of our lives … out on our community, for us to go out camping, hunting.

This in turn has cascading health effects as people 'stay in community and live off the roadhouse'. When asked when the difference started to be noticeable Nodea referred to his experiences growing up in Warmun 'we had the weather system, rain in late October, November, December, but it is now later'.

Nodea defined climate change as an issue within Ngarranggarni, identifying climate change as an imbalance that needed to be rectified by going 'back to Gija culture and practice to control this climate change. [The] only way [is to] go back on the land'. He argues that 'nothing much [is] happening out there due to climate change' so the answer was to 'get back to the land. Talk to the land and the land talks back to you and Gija Old People understand the way the land talks back to you'.

According to Nodea, this is also an intergenerational issue. Younger generation finds it very hard to understand because we are 'out of sight out of mind' adding that:

> Old People have belief in land and talking to the land – land talks back. We say there's nothing out there. Old People say there's nothing out there because you know nothing. In other words, you have to understand – to look after that knowledge of the land. … young people find it very hard to understand because Old People have more knowledge of the land than anybody. They know what is right for us. So in terms of 'primary' time they can predict the weather and changes, but we young people we just trust … [and ask] what's going on?

Two-way learning

Two-way learning was implemented at Warmun when the school commenced in the late 1970s at the instigation of the Old People. This over three decades before the United Nations University Traditional Knowledge Initiative advocated for 'bothways' education as a way of ensuring that intergenerational knowledge that strengthens community resilience is passed on across generations, noting:

> In many countries, formal education continues to contribute to the erosion of indigenous languages and knowledge. Compulsory schooling not only interrupts traditional channels of cultural transmission by removing children daily from family and community settings, but also inculcates children with external values that may clash with and even undermine traditional teachings. This loss of indigenous language and knowledge weakens the social capital of younger generations, which may result in diminished survival skills.[26]

Across Australia, Indigenous art centres have emerged as places where traditional language, culture and law are preserved, sustained and passed to the next generation through the production of contemporary art. This art is sold across the world, producing income for artists and their communities. In this respect, it plays an ambassadorial and educational role across the globe, serving as a bridge between senior Indigenous knowledge, and a world that is interested in knowing more about Indigenous knowledge. Increasing interest in this knowledge has extended into questions of how to address the challenges of climate change.

An independent review of Australia's Environmental Protection and Biodiversity Act undertaken in 2020 by Professor Graham Samuel, who is an expert in public policy in economic reform and regulation, argued for reforms that, amongst others, harnessed 'the knowledge of Indigenous Australians to better inform how the environment is managed'.[27] The Samuel Report noted that:

> The operation of the EPBC Act has failed to harness the extraordinary value of Indigenous knowledge systems that have supported healthy Country for over 60,000 years in Australia. A significant shift in attitude is required, so that we stop, listen and learn from Indigenous Australians and enable them to effectively participate in decision-making. National-level protection of the cultural heritage of Indigenous Australians is a long way out of step with community expectations. As a nation, we must do better.[28]

The review took submissions from 30,142 people and organisations and recommended 'a pathway of staged reform', with a number that were identified as 'Indigenous-specific'. These Indigenous-specific reforms included reviewing national-level cultural heritage protections and co-designing of those reforms identified as important to Indigenous Australians; incorporating Indigenous knowledge and science into decision-making; developing a 'National Environmental Standard' that would ensure Indigenous engagement and participation in decision-making, and refine it 'through an Indigenous-led process'; and '[w]ork with Indigenous Australians to meet their aspirations to manage their land in partnership with the Commonwealth'.[29] The Report emphasised the importance of 'sustained engagement' with Indigenous Australians, and the need to 'codesign reforms that are important to them'.[30]

The Intergovernmental Panel on Climate Change (IPCC) is the United Nations body for assessing the science related to climate change.[31] The IPCC brings together expert author panels to prepare reports on various aspects of climate change. Like the Australian review of the Environmental Protection Act, the IPCC has also acknowledged the role of Indigenous knowledge in research on, and understanding of, climate change. In March 2021 the virtual meeting of Working Group II, which assesses climate change impacts, adaptation and vulnerability, was hosted by Guatemala. The IPCC Chair, Hoesung Lee, opened the meeting with a statement that:

> IPCC assessments draw on and respect indigenous knowledge all over the world, and that is particularly the case for this group for the assessments of Working Group II. We recognise that:

- Traditional agriculture practices can be resilient to climate and non-climate stressors.
- And indigenous knowledge systems and practices allow local people to adapt to many climatic changes.
- Such initiatives based on indigenous knowledge have a foundation of legitimacy in local communities.[32]

This meeting of the Working Group members included 'a presentation by a Mayan leader on the relevance of indigenous knowledge'.[33]

A year earlier, the 52nd Session of the IPCC held in Paris in February 2020, and titled 'Planet in Peril: Transforming the Course of Climate Action', brought together

> eminent scientists, climate experts, youth leaders and representatives of indigenous peoples to discuss and share views on the ways and means to raise ambition for more effective climate action and the role and contributions of science and knowledge.[34]

The 'Summary for Policymakers' of the *IPCC Special Report on the Ocean and Cryosphere in a Changing Climate* discussed the 'benefits of combining scientific with local and indigenous knowledge to develop suitable options to manage climate change risks and enhance resilience'.[35] It noted the impact of the climate crisis on community health, economics and cultural preservation, stating that:

> Climate change impacts on marine ecosystems and their services put key cultural dimensions of lives and livelihoods at risk … This includes potentially rapid and irreversible loss of culture and local knowledge and Indigenous knowledge, and negative impacts on traditional diets and food security, aesthetic aspects, and marine recreational activities (B.8.4).[36]

The language barrier

The 2019 IPCC 'Summary Report for Policymakers' argued for the inclusion of Indigenous knowledge and local knowledge in climate studies, suggesting that multiscale linkages between institutions, local organisation and Indigenous communities would 'benefit adaptation' (C.4.2), and identifying the need for a form of climate literacy that would build 'public awareness, understanding and social learning about locality-specific risk and response potential' (C.4.4).[37]

Principles of inclusion, co-design, knowledge sharing, shared decision-making and Indigenous-led process are worthwhile aims, but devising a process that facilitates effective co-design, the sharing of knowledge and decision-making, and enables Indigenous-led processes is less straightforward. While it is clear that the stakes are extremely high and having a seat at the table is important for Indigenous communities, it is less clear how invitations to participate are extended and how this process enables, or disenables, input from Indigenous people. As Minnie Degawan, director of Conservation International's Indigenous & Traditional Peoples Program (ITPP), noted

> reliance on published materials is furthering the exclusion of Indigenous Peoples in the important discussions going on today regarding solutions to the climate crises and biodiversity loss. A huge body of important knowledge is discarded for this simple reason. If we are to truly be inclusive in finding solutions for climate change and biodiversity loss, we should not discriminate against different knowledge systems, such as those of Indigenous Peoples, simply because these are shared differently.[38]

The questions of how senior knowledge becomes available, and how the knowledge from Senior Knowledge holders can become embedded in institutional and organisation processes, are not straightforward. Of the 7,000 or so languages spoken globally, around 2,300 are spoken in Asia. In the Philippines, 180 languages are spoken by around 108 million people. In Papua New Guinea, 850 languages are spoken by a population of around 7.6 million people. Timor Leste has some 15–20 languages in a population of around 1.3 million people. While these figures illustrate some of the complexity of bringing diverse knowledge to international forums there are much more significant issues. Many languages are now only spoken by Senior Knowledge holders who hold the deep and located knowledge needed to contribute to the climate change challenge. In addition, many forms of language include 'high' language, which is used by Senior Knowledge holders for describing sacred/secret, ceremonial or complex high-level forms of knowledge. For example, amongst Tetun speakers in Timor Leste ritual Tetun Terik 'is an oral register which is controlled by traditional ritual specialists. It uses 'poetic parallelism, specialist vocabulary and pervasive metaphor' in formal ritual contexts and '[m]any of the terms are never used in daily speech, and are poorly understood by the bulk of the population'.[39] These 'high' forms of language are not available to all members of the community and can be used to screen meaning from younger community members who have not yet earned the right to access aspects of ceremonial or legal knowledge.

In Australia, there are 300 distinct Indigenous languages, more than 700 named language varieties, with around 15 of these being acquired by children as their first language. Around 80 or so of these are still spoken by elderly speakers, and there are a few additional languages in revival.[40] Their roots stretch back to the time when Australia was first occupied some 65,000 years or more ago. Each language explains, and constructs a world view, that is different and embedded in the environment and the society where they are located, and from which they grew. They encapsulate deep knowledge that is not easily translated, and even when translation is available it is out of context making it difficult to contextualise meaning. For example, the Bininj Gun-wok people of Arnhem Land in northern Australia have several different words that refer to the English word 'hop'. There is worrowkme, the hop of a person; djarrwahme, the hop of the female antilopine kangaroo; mawudme, the hop of male antilopine kangaroo; muribardme, the hop of the female euro, black rock kangaroo; and njedjme, the hop of male wallaroo. Each word brings together knowledge of species identification, gendered relationships and customary responsibilities. Things get more complex, however, when complex concepts are held within one word. Take for example the English translation of the Bininj Gun-wok word burrmaarlarl, which is

> the name for a male antilopine kangaroo that is resting, lying on its side in part shade during the heat of the day and the movement of the dappled light on its fur gives it the appearance of having covered itself with white clay, just as hunters do when they hunt these large animals.[41]

The idea of a United Nations translator providing an accurate translation of a sentence that contains three such words presents a fascinating prospect. Capturing this complex, senior knowledge in a roundtable forum in Paris, or through a zoom meeting with five-minute representative presentations (as was the case with the G20 Culture Ministers' Meeting in Rome in July 2021) seems particularly challenging.[42] There are, however, many records, produced by Indigenous communities that share Indigenous knowledge and ideas about climate change.[43] For example, the Warmun Art Centre published a book *Jadagen Warnkan Barnden: Wet Time Cold Time Hot Time Changing Climate in Gija Country* which presented artwork and accompanying accounts of how Gija people measure and understand climate and climate change.[44]

Principle for Indigenous-led climate change research

The ability to guarantee co-design, knowledge-sharing, shared decision-making and Indigenous-led processes that properly enable Senior Indigenous Knowledge to drive research, forums and programs is not a simple task. Returning to first principles seems a good place to start. First, there is a clear need to identify those gaps where Indigenous knowledge is not, and does not have the opportunity to be included. Second, there is a parallel need to support Indigenous leaders and their communities in their efforts to preserve knowledge so that it is available into the future. Both require multi-faceted approaches that are cross-cultural, equitable, interdisciplinary, flexible, nuanced, rigorous and properly resourced.[45]

As this chapter has discussed, much valuable knowledge about climate is part of what is obliquely referred to as 'intangible heritage'. Indigenous institutions and knowledge holders have structural mechanisms that ensure 'intangible' knowledge is protected, preserved and can be transmitted in ways that are effective and appropriate. Mechanisms for acknowledging and protecting Indigenous Knowledge are enshrined in both the 2003 UNESCO Convention for the Safeguarding of the Intangible Cultural Heritage and the United Nations Declaration on the Rights of Indigenous Peoples (2007).

In Australia, the AIATSIS Code of Ethics for Aboriginal and Torres Strait Islander Research governs research undertaken in partnership with Indigenous people and communities.[46] The Code has four principles, which serve as the basis for effective co-design, proper sharing of knowledge and decision-making, and enabling Indigenous-led processes. Some of the fundamental concerns of these principles are included below, and are:

PRINCIPLE 1: Indigenous self-determination
: This requires that research projects that involve Aboriginal and Torres Strait Islander communities should 'be underpinned by a negotiated agreement or protocol'. Such documents should clearly define roles and responsibilities. They should be clear that there is an understanding of how the research is to be undertaken. They should provide 'greater recognition of and protection for the rights of Indigenous peoples' by extending current (limited) legal requirements for recognising and protecting Indigenous cultural and intellectual property. Finally, self-determination requires free, prior and informed consent 'through culturally and linguistically appropriate mechanisms'.

PRINCIPLE 2: Indigenous leadership
: This principle requires that Aboriginal and Torres Strait Islander research is led by Indigenous people. This means that research is 'informed by Indigenous priorities, values, perspectives and voices', and that Aboriginal and/or Torres Strait Islander people have a genuine (and transparent) role in decision-making and leadership. This principle should be evident in the life cycle of the program 'from conceptualisation to communication of research'.

PRINCIPLE 3: Impact and value
: Adherence to this principle ensures the benefit, impact and value of Aboriginal and Torres Strait Islander research is available to Indigenous people, and that there is a full understanding of the risks as well the benefit by all parties.

PRINCIPLE 4: Sustainability and accountability
: Sustainable research programs are environmentally, culturally, socially and economically sustainable. Such programs also ensure that knowledge and data collected remains available 'for use by current and future generations, for example through the return of materials to communities and through appropriate archiving'.

Many countries have similar ethical frameworks for working with Indigenous communities as do a number of professional organisations and institutions. Together these form a rich resource to support cultural heritage experts as they traverse the complicated and critically significant domain of Indigenous Knowledge about climate, climate change and climate crisis. This chapter has shown that, while there is considerable affective work being done across the sectors, the challenge, it seems, is to ensure that there are the resources required to enable the scope of research necessary for Indigenous knowledge to be a critical and central part of the climate crisis response. Indigenous Knowledge is central to the research described in the following chapter which focuses on the information that can be derived from rock art, and the risks to these sites from the impacts of climate change.

Notes

1 UNESCO 2021a, 'Indigenous knowledge and climate change', https://en.unesco.org/links/climatechange. Note that the term 'Indigenous' appears both with and without a capital 'I' in the literature. In this book, Indigenous is used with a capital 'I' when referring to people (as for European, Australian, Asian, French and the like), and with a lower case 'i' when referring to indigenous flora and fauna.
2 Van den Bergh, GD, Li, B, Brumm, A et al 2016, 'Earliest hominin occupation of Sulawesi, Indonesia', *Nature*, Vol. 529, pp. 208–11.
3 Brumm, A, Ririmasse, M, Adhityatama et al 2021, 'Oldest cave art found in Sulawesi', *Science Advances*, Vol. 7, No. 3, pp. 1–12.
4 Clarkson, C, Jacobs, Z, Marwick, B et al 2017, 'Human occupation of northern Australia by 65,000 years ago', *Nature,* Vol. 547, pp. 306–10; and Florin, SA, Roberts, P, Marwick, B et al 2021, 'Pandanus nutshell generates a palaeoprecipitation record for human occupation at Madjedbebe, northern Australia', *Nature Ecology & Evolution,* Vol. 5, pp. 295–303.
5 Finch, D, Gleadow, A, Hergt, J et al 2021, 'Ages for Australia's oldest rock paintings', *Nature Human Behaviour,* Vol. 5, pp. 310–18.
6 UNESCOb, 2021, 'Indigenous knowledge and knowledge transmission, revitalizing knowledge - Indigenous education', https://en.unesco.org/links/transmission
7 Handayani, RD, Wilujeng, I, Zuhdan, K & Prasetyo, ZK 2018, 'Elaborating Indigenous knowledge in the science curriculum for the cultural sustainability', *Journal of Teacher Education for Sustainability,* Vol. 20, No. 2, p. 75.
8 Bureau of Meteorology (BOM) 2016, 'Indigenous weather knowledge', http://www.bom.gov.au/iwk/
9 Cullen-Unsworth, L & Maclean, K 2015, ''We now have a name for some of the big changes happening on our Bubu [country]'. The role of Indigenous knowledge for the management of cultural landscapes in a changing climate: a case study of the Kuku Nyungkal people from the Queensland Wet Tropics, Australia' in D Harvey & J Perry (eds) *The Future of Heritage as Climates Change: Loss, Adaptation and Creativity,* Taylor & Francis Group, Oxon, p. 96.
10 Clarke, PA 2009, 'Australian Aboriginal ethnometeorology and seasonal calendars', *History and Anthropology*, Vol 20, No. 2, p. 80.
11 Clarke 2009, p. 80.
12 Clarke 2009, p. 81.
13 BOM 2015, 'Language, culture and environmental knowledge', *Indigenous Weather Knowledge,* http://www.bom.gov.au/iwk/culture.shtml
14 The Australian Bureau of Meteorology established the Indigenous Weather Knowledge website in 2002 in collaboration with the Aboriginal and Torres Strait Islander Commission (ATSIC) and Monash University's Centre for Indigenous Studies.
15 Smith, B & Parke, E 2021, 'Oldest-known rock art in Australia is 17,300-year-old kangaroo in the Kimberley, wasp nests show', *ABC News*, 23 February 2021.
16 Western Australian Government Department of Sustainability, Environment, Water, Population and Communities 2011, 'The West Kimberley, Western Australia', Australian Government Department of Agriculture, Water & the Environment, https://www.awe.gov.au/parks-heritage/heritage/publications/west-kimberley/wk-factsheet

17 McConnell, K 1998, 'The prehistoric use of Chenopodiaceae in Australia: evidence from carpenter's gap shelter 1 in the Kimberley, Australia', *Vegetation History and Archaeobotany*, Vol. 7, pp. 179–88; University of Western Australia 2019, 'New site reveals 50,000 years of Aboriginal occupation in Kimberley', *UWA News*, 21 August, https://www.news.uwa.edu.au/archive/2019082111558/regional/new-site-reveals-50000-years-aboriginal-occupation-kimberley/
18 Carrington, S, Kimberley, J, Kowalski, V et al 2014, 'Conservation and the production of shared knowledge' in J Bridgland (ed) *ICOM-CC 17th Triennial Conference Preprints, Melbourne, 15–19 September 2014*, International Council of Museums, Paris, pp. 1–8.
19 Carrington et al 2014.
20 Carrington et al 2014.
21 Warmun Art Centre, 'About Warmun art', https://warmunart.com.au/about/
22 Barney, G, Cann, C, Carrington, B et al 2013, *Jadagen Warnkan Barnden: Changing Climate in Gija Country,* Warmun Art Centre, Warmun, NT, Australia.
23 Purdie, S 2020, 'Ngarranggarni – Ngalim Ngalimbooroo Ngagenybe' [exhibition], National Portrait Gallery, https://www.portrait.gov.au/content/shirley-purdie-ngarranggarni-row3
24 Purdie 2020.
25 Kofod, F 2013, 'Language and the land' in G Barney, C Cann & B Carrington et al 2013, *Jadagen Warnkan Barnden: Changing Climate in Gija Country,* Warmun Art Centre, Warmun, NT, Australia, p. 11.
26 Nakashima, D, Galloway McLean, K, Thulsrup, H, Ramos-Castillo, A & Rubis, J 2012, *Weathering Uncertainty: Traditional Knowledge for Climate Change Assessment and Adaptation*, UNESCO, Paris & Darwin, Australia.
27 Samuel, G 2020, *Independent Review of the EPBC Act – Final Report*, Department of Agriculture, Water and the Environment, Canberra, p. iii.
28 Samuel 2020, p. iii.
29 Samuel 2020, p. 193.
30 Samuel 2020, p. 195.
31 IPCC 2021a, 'Intergovernmental panel on climate change', https://www.ipcc.ch/
32 IPCC 2021c, 'Statement by IPCC Chair: opening of the fourth lead author meeting of Working Group II for the sixth assessment report', *IPCC NewsRoom*, 1 March, https://www.ipcc.ch/2021/03/01/opening-statement-by-ipcc-chair-wgii-ar6-lam4/
33 IPCC 2021b, 'IPCC authors meet virtually to work on upcoming report', *IPCC NewsRoom*, 25 February, https://www.ipcc.ch/2021/02/25/press-release-wgii-ar6-fourth-lead-author-meeting/
34 IPCC 2020, 'IPCC meets in Paris to discuss major 2022 climate change report, other business', *IPCC NewsRoom*, 20 February, https://www.ipcc.ch/2020/02/20/ipcc-meets-in-paris-to-discuss-major-2022-climate-change-report-other-business/
35 IPCC 2019b, 'Choices made now are critical for the future of our ocean and cryosphere', *IPCC NewsRoom*, 25 September, https://www.ipcc.ch/2019/09/25/srocc-press-release/
36 IPCC 2019a, 'Summary for policymakers' in H-O Pörtner, DC Roberts, V Masson-Delmotte, P Zhai, M Tignor, E Poloczanska, K Mintenbeck, A Alegría, M Nicolai, A Okem, J Petzold, B Rama & NM Weyer (eds) *IPCC Special Report on the Ocean and Cryosphere in a Changing Climate*, in press.
37 IPCC 2019a.
38 Degawan, M 2020, 'Subtle ways of excluding Indigenous voices', *International Union for the Conservation of Nature (IUCN)*, 11 October, https://www.iucn.org/news/commission-environmental-economic-and-social-policy/202010/subtle-ways-excluding-indigenous-voices-0
39 Williams-Van Klinken, C 2001, 'High registers of Tetun Dili: Portuguese press and purist priests', *Proceedings of the 2001 Conference of the Australian Linguistics Society*, p. 2, http://www.als.asn.au/proceedings/als2001/williams-vanklinken.pdf

40 Nordlinger, R 2018, 'The genius of Australian Indigenous languages, and why they are important for all of us', Public Lecture, The University of Melbourne, 30 May, from 55:51, https://vimeo.com/273614736
41 Nordlinger 2018.
42 MiC_Italia 2021, 'G20 Cultural Ministerial', G20 Italia 2021, YouTube video, 12 April, https://www.youtube.com/watch?v=5h2tyWCPOLQ
43 Leonard, S, Parsons, M, Olawsky, K & Kofod, F 2013, 'The role of culture and traditional knowledge in climate change adaptation: insights from East Kimberley, Australia', *Global Environmental Change*, Vol. 23, No. 3, pp. 623–32.
44 Barney et al 2013.
45 In this respect the 'Evaluation of UNESCO's action to revitalize and promote Indigenous languages: within the framework of the International Year of Indigenous Languages', March 2021IOS/EVS/PI/194, https://unesdoc.unesco.org/ark:/48223/pf0000376719.locale=fr noted that a particular challenge had been that 'UNESCO was asked to lead this effort within existing resources and relying on a very small core team … without a budget for intersectoral activities, collaboration between sectors was limited to information sharing and activities in Africa and the Arab States were few'.
46 Australian Institute of Aboriginal and Torres Strait Islander Studies (AIATSIS) 2020, *AIATSIS Code of Ethics for Aboriginal and Torres Strait Islander Research,* AIATSIS, Canberra.

References

Australian Institute of Aboriginal and Torres Strait Islander Studies (AIATSIS) 2020, *AIATSIS Code of Ethics for Aboriginal and Torres Strait Islander Research*, AIATSIS, Canberra.

Barney, G, Cann, C, Carrington, B, Juli, M, Nodea, N, Nyadbi, L, Peters, R, Purdie, S, Thomas, M, Kofod, F, Hunt, A & Leonard, S 2013, *Jadagen Warnkan Barnden: Changing Climate in Gija Country*, Warmun Art Centre, Warmun, NT, Australia.

Brumm, A, Ririmasse, M, Adhityatama, S, Oktaviana, AA, Burhan, B, Hakim, B, Lebe, R, Zhao, JX, Sulistyarto, P, Sumantri, I & Aubert, M 2021, 'Oldest cave art found in Sulawesi', *Science Advances*, Vol. 7, No. 3, pp. 1–12, doi:10.1126/sciadv.abd4648

Bureau of Meteorology (BOM) 2015, 'Language, culture and environmental knowledge', *Indigenous Weather Knowledge*, http://www.bom.gov.au/iwk/culture.shtml

BOM 2016, 'Indigenous weather knowledge', http://www.bom.gov.au/iwk/

Carrington, S, Kimberley, J, Kowalski, V, Lewincamp, S, Mung Mung, P, Nodea, G, Park, R, Scott, M & Sloggett, R 2014, 'Conservation and the production of shared knowledge' in J Bridgland (ed) *ICOM-CC 17th Triennial Conference Preprints, Melbourne, 15–19 September 2014*, International Council of Museums, Paris, pp. 1–8.

Clarke PA 2009, 'Australian Aboriginal ethnometeorology and seasonal calendars', *History and Anthropology*, Vol 20, No. 2, pp. 79–106, doi:10.1080/02757200902867677

Clarkson, C, Jacobs, Z, Marwick, B, Fullagar, R, Wallis, L, Smith, M … Pardoe, C 2017, 'Human occupation of northern Australia by 65,000 years ago', *Nature*, Vol. 547, pp. 306–310, doi:10.1038/nature22968

Cullen-Unsworth, L & Maclean, K 2015, "We now have a name for some of the big changes happening on our Bubu [country]'. The role of Indigenous knowledge for the management of cultural landscapes in a changing climate: a case study of the Kuku Nyungkal people from the Queensland Wet Tropics, Australia' in D Harvey & J Perry (eds) *The Future of Heritage as Climates Change: Loss, Adaptation and Creativity*, Taylor & Francis Group, Oxon, pp. 95–110.

Degawan, M 2020, 'Subtle ways of excluding Indigenous voices', *International Union for the Conservation of Nature (IUCN)*, 11 October, https://www.iucn.org/news/commission-environmental-economic-and-social-policy/202010/subtle-ways-excluding-indigenous-voices-0

Finch, D, Gleadow, A, Hergt, J … Levchenko, VA 2021, 'Ages for Australia's oldest rock paintings', *Nature Human Behaviour*, Vol. 5, pp. 310–318, doi:10.1038/s41562-020-01041-0

Florin, SA, Roberts, P, Marwick, B, Patton, NR, Shulmeister, J, Lovelock, CE … Clarkson, C 2021, 'Pandanus nutshell generates a palaeoprecipitation record for human occupation at Madjedbebe, northern Australia', *Nature Ecology & Evolution*, Vol. 5, pp. 295–303, doi:10.1038/s41559-020-01379-8

Handayani, RD, Wilujeng, I, Zuhdan, K & Prasetyo, ZK 2018, 'Elaborating Indigenous knowledge in the science curriculum for the cultural sustainability', *Journal of Teacher Education for Sustainability*, Vol. 20, No. 2, pp. 74–88.

IPCC 2019a, 'Summary for policymakers' in H-O Pörtner, DC Roberts, V Masson-Delmotte, P Zhai, M Tignor, E Poloczanska, K Mintenbeck, A Alegría, M Nicolai, A Okem, J Petzold, B Rama & NM Weyer (eds) *IPCC Special Report on the Ocean and Cryosphere in a Changing Climate*, in press.

IPCC 2019b, 'Choices made now are critical for the future of our ocean and cryosphere', *IPCC NewsRoom*, 25 September, https://www.ipcc.ch/2019/09/25/srocc-press-release/

IPCC 2020, 'IPCC meets in Paris to discuss major 2022 climate change report, other business', *IPCC NewsRoom*, 20 February, https://www.ipcc.ch/2020/02/20/ipcc-meets-in-paris-to-discuss-major-2022-climate-change-report-other-business/

IPCC 2021a, 'Intergovernmental panel on climate change', https://www.ipcc.ch/

IPCC 2021b, 'IPCC authors meet virtually to work on upcoming report', *IPCC NewsRoom*, 25 February, https://www.ipcc.ch/2021/02/25/press-release-wgii-ar6-fourth-lead-author-meeting/

IPCC 2021c, 'Statement by IPCC Chair: opening of the fourth lead author meeting of Working Group II for the sixth assessment report', *IPCC NewsRoom*, 1 March, https://www.ipcc.ch/2021/03/01/opening-statement-by-ipcc-chair-wgii-ar6-lam4/

Kofod, F 2013, 'Language and the land', in G Barney, C Cann, B Carrington, M Juli, N Nodea, L Nyadbi, R Peters, S Purdie, M Thomas, F Kofod, A Hunt & S Leonard (eds) *Jadagen Warnkan Barnden: Changing Climate in Gija Country*, Warmun Art Centre, Warmun, NT, Australia, p. 11.

Leonard, S, Parsons, M, Olawsky, K & Kofod, F 2013, 'The role of culture and traditional knowledge in climate change adaptation: insights from East Kimberley, Australia', *Global Environmental Change*, Vol. 23, No. 3, pp. 623–632, doi:10.1016/j.gloenvcha.2013.02.012

McConnell, K 1998, 'The prehistoric use of Chenopodiaceae in Australia: evidence from Carpenter's gap shelter 1 in the Kimberley, Australia', *Vegetation History and Archaeobotany*, Vol. 7, pp. 179–188.

MiC_Italia 2021, 'G20 cultural ministerial', G20 Italia 2021, YouTube video, 12 April, https://www.youtube.com/watch?v=5h2tyWCPOLQ

Nakashima, D, Galloway McLean, K, Thulsrup, H, Ramos-Castillo, A & Rubis, J 2012, *Weathering Uncertainty: Traditional Knowledge for Climate Change Assessment and Adaptation*, UNESCO, Paris & Darwin, Australia.

Nordlinger, R 2018, 'The genius of Australian Indigenous languages, and why they are important for all of us', public lecture, The University of Melbourne, 30 May, https://vimeo.com/273614736

Purdie, S 2020, 'Ngarranggarni – Ngalim Ngalimbooroo Ngagenybe' [exhibition], National Portrait Gallery, https://www.portrait.gov.au/content/shirley-purdie-ngarranggarni-row3

Samuel, G 2020, *Independent Review of the EPBC Act – Final Report*, Department of Agriculture, Water and the Environment, Canberra.

Smith, B & Parke E 2021, 'Oldest-Known rock art in Australia is 17,300-year-old kangaroo in the Kimberley, wasp nests show', ABC News, 23 February 2021, https://www.abc.net.au/news/science/2021-02-23/oldest-rock-art-australia-kimberley-kangaroo-mud-wasp-nest/13173792

UNESCO, 2021a, 'Indigenous knowledge and climate change', https://en.unesco.org/links/climatechange

UNESCO, 2021b, 'Indigenous knowledge and knowledge transmission, revitalizing knowledge - Indigenous education', https://en.unesco.org/links/transmission

UNESCO, 2021c, 'Evaluation of UNESCO's action to revitalize and promote Indigenous languages: within the framework of the International Year of Indigenous Languages' March 2021IOS/EVS/PI/194, https://unesdoc.unesco.org/ark:/48223/pf0000376719.locale=fr

University of Western Australia 2019, 'New site reveals 50,000 years of Aboriginal occupation in Kimberley', *UWA News*, 21 August, https://www.news.uwa.edu.au/archive/2019082111558/regional/new-site-reveals-50000-years-aboriginal-occupation-kimberley/

Van den Bergh, GD, Li, B, Brumm, A, Grün, R, Yurnaldi, D, Moore, MW, Kurniawan, I, Setiawan, R, Aziz, F, Roberts, RG, Suyono, Storey, M, Setiabudi, E & Morwood, MJ 2016, 'Earliest hominin occupation of Sulawesi, Indonesia', *Nature*, Vol. 529, pp. 208–211, doi:10.1038/nature16448

Warmun Art Centre, 'About Warmun art', https://warmunart.com.au/about/.

Western Australian Government Department of Sustainability, Environment, Water, Population and Communities 2011, 'The West Kimberley, Western Australia', Australian Government Department of Agriculture, Water & the Environment, https://www.awe.gov.au/parks-heritage/heritage/publications/west-kimberley/wk-factsheet

Williams-Van Klinken, C 2001, 'High registers of Tetun Dili: Portuguese press and purist priests', *Proceedings of the 2001 Conference of the Australian Linguistics Society*, http://www.als.asn.au/proceedings/als2001/williams-vanklinken.pdf

4

CLIMATE CHANGE AND ROCK ART: A VALUABLE RESOURCE AT RISK

Marcelle Scott and Robyn Sloggett

Rock art is a valuable resource: as historical documentation, and as a tool for exploring climate change, species extinction and changed land use. Rock art is also a record of continuity and change in current contemporary practices. It is a window into agricultural practice and the use of species; a way to explore social practices in food production; and finally, it provides data to contest assumptions about definitions such as 'hunter/gatherer' or 'agrarian practice', that have arisen from a linear and colonial evolutionary view of the progress of societies. The intersection of local knowledge, historic documents, place, memory and the adaptation and application of tradition knowledge are all important parts of the story of the relationship between heritage and climate change. This makes the conservation of rock art a particularly urgent issue as the impacts of climate change threaten to disrupt and even destroy the habitat and environment surrounding these significant place-based heritage documents. Assessing, documenting and understanding deterioration mechanisms, and rapid onset threats to rock art, also help in understanding the local impacts of climate change. This chapter explores these points and examines what are the major concerns in preserving rock art in the face of the environmental threats precipitated by the climate crisis.

Rock art: the oldest form of documentary heritage

It's hard to think of a portrait of the Warty Pig in terms of great art. The adult male Sulawesi Warty Pig, or more properly, *Sus celebensis*, might be considered unattractive, except perhaps to another Sulawesi Warty Pig. Its visage is distinguished by the three pairs of facial warts. The largest, the pair of preorbital warts, are located at the front of the eye socket, the smaller infraorbital warts run along the front of the pig's face, and the mandibular warts sit along the lower jaw and are contained within whorls of hair. It's the kind of face only a mother can love. The

DOI: 10.4324/9781003163312-5

Sulawesi Warty Pig is not small. Its body is generally between 80 and 130 cm in length, but some of the larger males can grow to an immense 160 cm from nose to the end of rump, stand 70 cm at the shoulder, and weigh in at a hefty 40–70 kgs. Some boars grow to a massive 150 kgs. The species is domesticated across eastern Indonesia, Timor Leste, New Guinea and parts of the Philippines, where it is a major food source and provides other resources such as ivory-like tusks. Given its lack of aesthetic endowments, the Warty Pig does not seem a likely candidate for one of the world's most significant paintings. Nevertheless, it is, for depicted in the rock art complex of Leang Tedongnge in the limestone karsts of Maros-Pangkep in South Sulawesi in Indonesia is most likely the earliest known dated artwork, a painting of a Sulawesi Warty Pig, with a firmly established minimum age of 45,500 years (Fig. 4.1).[1]

Not all of the species depicted in rock art have survived as long as the Warty Pig. The image of a giant thunderbird, *Genyornis newton*, has been reported in a rock art site in western Arnhem Land, in northern Australia.[2] At around 200–300 km, the *Genyornis newtoni* is a weightier proposition than the Warty Pig. *Genyornis newtoni* is now extinct, thought to have declined in line with the decline of its primary habitat, dry scleromorphic shrubland. The decline of the scleromorphic shrubland accelerated during the Pleistocene Epoch, around 65,000 years ago, at around the same time as humans arrived in Australia. Previously it was thought *Genyornis* had not survived beyond 47,000 BP. The rock art depiction of *Genyornis newtoni* has been dated to not before 13,739 to 13,976 BP,[3] so the image of a *Genyornis newtoni* in a painting made after this date raises questions about extinction dating sequences, the possibility of remnant populations and the dating of rock art. The giant echidna,

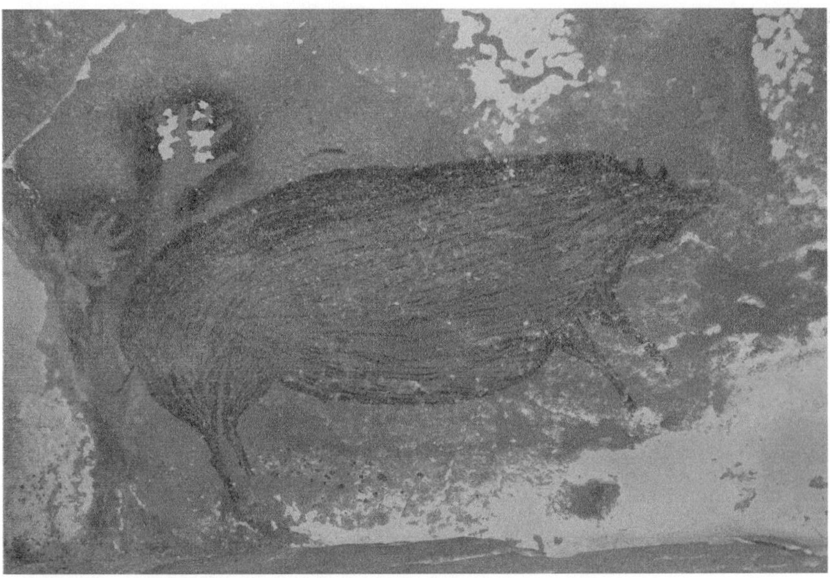

FIGURE 4.1 Image of the Sulawesi Warty Pig cave art. Photo by Maxime Aubert

Zaglossus sp., which survived to around 18,000 BP, also appears in rock art in the Arnhem Land plateau.[4] There is also an image of what is possibly a sthenurine kangaroo, *Sthenurus stirlingi*, a giant macropod about 3–6 times larger than current macropod species[5] and which survived during the Late Pleistocene epoch (around 129,000 to 11,700 years ago).[6] It is also generally accepted that rock art in Arnhem Land depicts the extinct thylacine, *Thylacinus cynocephalus*, otherwise known as the Tasmanian Tiger.[7] Becoming extinct on mainland Australia about 4,000 years ago, the Tasmanian Tiger survived in Tasmania until the 1930s. As Paul Taçon and Steve Webb note, using images of rock art as evidence of megafauna is contentious, 'there are extreme views varying from full acceptance that there are surviving rock art depictions of megafauna to full rejection'.[8] Nevertheless, such heritage forms an important part of attempts to understand climate change and species extinction in the past.

As a form of historical Indigenous documentation, rock art provides alternative views to those contained in written historical documents. For example, the trepang trade between northern Australia and China is well-documented. Written accounts record how praus (perahu/ship) from Macassa came to Australia to trade with the Aboriginal people who lived along Malege (Northern Territory coast) and Kayu Djawa (Kimberley coast).[9] Contemporary Aboriginal oral and ceremonial traditions preserve the history of the Macassan traders, and rock art images record the trade. On the traditional lands of the Maung people at Djulirri, in the Wellington Range on the northernmost edge of the Arnhem Land Plateau, there are pictures of Macassan praus in rock art from the mid-seventeenth century. Based on accelerator mass spectrometry (AMS) analysis of a beeswax snake design overlaying a painting of one prau, archaeologist and anthropologist Dr Sally May suggests the date 1664, and possibly much earlier.[10]

May concludes that:

> The study of Macassan-related rock art in Australia is only at its beginning … [and] new, important insights into the nature of cross-cultural contact during the past few hundred years are likely to emerge. A related issue to explore is whether older rock art will reveal much earlier contact with people from across Asia … a whole new picture of Aboriginal Australian contact with Asia will finally refute the long-held theory that Aboriginal Australians were isolated from the rest of the world until just a few hundred years ago.[11]

Rock art: heritage for the future

There are at least 100,000 rock art sites in Australia, and new previously undocumented sites are being discovered and documented each year. While they are compelling for the links they provide to the past, their most salient cultural links are to the Indigenous communities and individuals who are the Traditional Owners of the place, many of whom are reclaiming the cultural significance of the stories associated with the specific sites.[12]

Archaeologist John Hayward compared rock art from Mirarr country in the northern Kakadu region of the Northern Territory dated to around 10,000 years ago, with cultural items from the same region held in the South Australian Museum. In comparing the rock art paintings with museum artefacts, Hayward found the rock art paintings to be so accurate 'as to be identifiable as specific types found in most ethnographic collections around the country'.[13] He found that the rock art provided 'a fascinating insight into the lives of past Aboriginal cultures not usually accessible through other archaeological methods'.[14] Hayward also examined depictions of material culture on rock art and the relationship of this record to known climate change sequences. He found that the most recent paintings depicted a new range of 'objects such as didjeridu, goose-wing fans and watercraft … reflecting the newly established freshwater environment'.[15] Hayward also identified links between rock art and ontological developments in Mirrarr art from around 10,000 years ago. He found

> the emergence of new spear types with a diverse range of hafted tips that appeared in paintings after the boomerang became scarce, seems to coincide with previously unseen figure types such as Yam Figures, and mythological creatures such as the Rainbow Serpent.[16]

These barbed spears were identified as a particular type of wooden carved barbed spear known as 'marrgidjba' that was 'thought to have originated in Central Arnhem Land'.[17] Information on a barbed spear in the South Australia Museum, collected in 1900, identified it as a highly prized trade item. Two of these depicted in the rock art have more barbs than any found in ethnographic collections. Hayward explained how rock art can provide evidence for changes in the manufacture and use of cultural material, explaining that images of 'multi-pronged spears, trident spears and others with serrated tips', became less common in more recent paintings, and 'by the time of European contact were unknown or at least unrecorded; subsequently no equivalent types are found in ethnographic collections. Such evidence is unavailable in ethnographic or archival collections'.[18]

Rock art can also demonstrate continuity of cultural practice. The Injalak Arts and Craft Centre is located south of the Malarrak rock art site and not far from Djulirri where the paintings of the Macassan praus can be found. At Injalak contemporary art practice includes weaving a particular bag, kulungunjngunj, a practice that 'has been continued by contemporary artists from Injalak Arts and Craft Centre in Kunbalanja (Oenpelli)'.[19] Kulungunjngunj appear in rock art at Malarrak. They are worn around the chest and carry powerful objects, and give power and courage to the wearer.[20] Emily Miller argues that the depiction of fibre objects in rock art has been under-researched, and that more focussed research, across the globe, would enable archaeologists, anthropologists and others to better understand the types and extent of objects from earlier cultures.[21] Noting that theirs was a preliminary study, Noelene Cole and Alan Watchman examined organic material found in inorganic paint samples in the Laura region of north Queensland.[22] Their findings point to the use of macerated and broken fibres added to the paint to better bind the individual components of ochres and medium, as well as evidence of a range of materials employed as brushes or other implements. They argue that the analysis of organic material remnants found in rock art provides a way of better understanding Indigenous paint technology. The remnant fibres from the epiphytic orchid *Cymbidium canaliculatum* align with the use of the mucilage from macerated orchids in the production of bark paintings in the twentieth century. Cole and Watchman argue that the range of fibres found, including human hair, demonstrate

> the existence of a most interesting paint technology, with important implications not only for rock art conservation measures, but also for the study of the complex intellectual and physical processes which lay behind the creation of the remarkable system of Aboriginal art at Laura.[23]

Veth, Myers, Heaney and Ouzman explored agricultural practices as depicted in rock art, finding

> a primary record of long-term sophisticated physical and symbolic manipulation of plants that fits neither into the simplistic categories of 'foraging' or of 'agriculture' ... a society in which people actively chose not to pursue orthodox agriculture while according plants a central place in their lives.[24]

Across the globe, the climate crisis, and associated impacts from ocean warming, rising sea levels, coastal erosion, increased rainfall, extreme weather events and seasonal changes, is threatening the micro and macro environments that have helped to protect the valuable resource that is rock art.

The effects of climate change on rock art

In a study that included field work, sample analysis and the intergenerational experiential knowledge of local Indigenous experts, Huntley et al examined 11 sites in the Maros-Pangkep limestone karsts where the Sulawesi Warty Pig was painted. This area is in one of 'the world's most atmospherically dynamic region, the Australasian monsoon domain'.[25] They examined the exfoliation of the limestone surfaces caused by the cycles of wetting and drying, resulting in the expansion and contraction of salt crystals, which have affected the sites over many thousands of years. They concluded that increased flooding, as well as the 'seasonal moisture injected via monsoonal rains' being retained as 'standing water in the rice fields and aquaculture ponds of the region' has increased the rate of deterioration from cyclical surface exfoliation, 'with more panel loss from exfoliation over recent decades than at any other time in living memory'.[26]

> Although the Pleistocene rock art of Maros-Pangkep has existed throughout tens of thousands of years of climate variation in response to the earth's orbital geometry, there is no doubt that global carbon cycling and greenhouse gasses are now significantly amplifying climatic responses … hastening the deterioration of the unique, irreplaceable record of early human artistic culture in a little-understood region, one that continues to provide important insights into the culture of the first peoples of IMC [Indonesian Maritime Continent].[27]

A number of rock art sites in northern Australia are also at risk from climate change impacts including sea-level rise, more intense cyclones, storm surges, extreme and more frequent rainfall and coastal erosion.[28] Indigenous rangers caring for country in the World Heritage-listed Kakadu National Park and the Djelk Indigenous Protected Area in Northern Australia have cultural responsibilities that require them to protect sites in these areas. Carmichael reported that several Djelk Rangers reported 'unprecedented inundation and obliteration of riparian rock art as a result of extreme flooding'.[29] As one of the rangers described:

> There's more erosion from water, creek side, by bigger floods. Big floods are ruining paintings and making erosion … some of these are really old paintings. Even when paintings are high up, water going right up and ruining the painting.[30]

The most senior Djelk Ranger added that:

> Climate change is really quick for us and really shocking for us, especially real bush mob – country men … . When we are on country, we record things with our mind – you see water mark gone right up, ruining the painting.[31]

The Kakadu National Park Rangers also reported climate change impacts on rock art. As one of the rangers said:

> People worrying a lot; people worrying about the sea, but you got to worry about what's coming from on top [i.e., freshwater flooding from the escarpment] once you've got that high tide doubled up … . At Canon Hill in last big wet in 2006 a few art sites there, which have never had artwork go under, went underwater. This art never went underwater before. I did not see the water go up, but I saw the damage in the dry season; saw the level of the water on the rock itself and on the trees.[32]

Another ranger expressed his concerns about the physical and cultural impacts:

> We're seeing a lot of changes. The big one, that's the climate change! We're savvy, we know! We're trying to tell the balanda [non-Indigenous person]

> what's going to happen in the future … I'm worried about sites down South Alligator River … our ancestors are there … very important to us, we see it's all changing, very important to us, it's all changing.[33]

Australia and Indonesia are not alone in needing to address this crisis. In the Northumberland region in northern England, 6000-year-old sandstone petroglyphs are at an increased risk of weathering from the predicted warmer, windier weather and seasonably variable rainfall patterns.[34] Acknowledging the challenges of measuring surface changes on ancient sandstone that has been exposed to the weather for thousands of years, Giesen et al concluded that 'recent rapid environmental change has triggered significantly increased rates of deterioration' and that 'more change has occurred in the last 50 years than the previous 6000'.[35] This finding accords with AH Viles' study of future climate change impacts on stone deterioration.[36] Giesen et al were aided in their forecasting by the UK Climate Projection Model, UKVP09, which currently generates regularly updated climate models of predicted weather parameters including wind speeds, 'precipitation, temperature, humidity, cloud cover, air pressure, and wind patterns' through to 2099.[37] Giesen et al propose a four-step management and preservation strategy for outdoor stone sites, which identifies a process to:

- Calibrate field data to local environments;
- Study specific weathering conditions and the actual stone in situ ;
- Prioritise those requiring immediate attention;
- Develop integrated preservation and management strategies that include awareness raising and education.

One general finding from their research is that deterioration of the petroglyphs is relative to wind exposure and soil condition, a process consistent with deterioration of building stone, making their recommendations applicable to a range of outdoor heritage buildings and sites.

Managing rock art at risk

The painting of the Warty Pig in Leang Tedongnge has survived 1,137,500 human generations, but like much other rock art, it may not survive this generation.[38] There are, however, few studies that provide practical advice on how to protect rock art against the impacts of climate change.

In 2013, the IPCC's Fifth Assessment Report of the Intergovernmental Panel on Climate found that the global mean surface temperature had increased 'by between about 0.3 and 0.6°C since the late 19th century'[39] and that as a result 'Models project a sea level rise of 50 cm from the present to 2100'.[40]

A decade ago, as the IPCC was gathering data for its report, the Getty Conservation Institute was collecting data on rock art preservation as part of its six-year Southern African Rock Art Project (SARAP).

The Getty Southern African Rock Art Project and the African-Australian Exchange that followed culminated in a forum held in Kakadu National Park in northern Australia in 2014. Results were brought together in the publication *Rock Art: A Cultural Treasure at Risk – How Can We Protect the Valuable and Vulnerable Heritage of Rock Art* in 2015. This report highlighted both the global significance of rock art, and the risks to its survival. These risks included development pressures, graffiti and vandalism, poor tourist management and natural impacts. While immediate direct interventions could be brought to bear on the first three risks, natural impacts are more insidious. The report identified these as

- Rock weathering;
- Water;
- Dust and mineral accretions;
- Vegetation growth;
- Animal impacts, and;
- Major environmental events.[41]

All six of these impacts can be related to new weather patterns resulting from climatic changes, as temperatures, rainfall, winds, shifts in species growth and location, and environmental disasters change in intensity, type or habit. Three areas present particular concern. These are rising sea levels, increased wildfire activity and severity, and microorganisms and algae.

Rising sea levels

In 2014, climate scientists Ben Marzeion and Anders Levermann reported that over the next two millennia the levels of current global mean temperature would result in the loss of about 40 UNESCO World Heritage Sites (6%) from rising sea levels. Many of these are not yet properly documented. They also noted the damage from subsidence that can result from ground water depletion such as the contraction of artesian wells during drought, and drawing water at an industrial scale such as from hydrofracking.[42] While these findings are alarming, even more alarming is the fact that UNESCO World Heritage Listings do not yet contain any record of the vast amounts of highly significant rock art sites that are located in areas most at risk from rising sea levels. Part of this is the nature of the world heritage listing with a statistical deficit of listings related to rock art across the Asia-Pacific region, but part of the problem also rests with the lack of documentation of rock art sites. Sulawesi's Maros-Pangkep sites, for example, are not part of the UNESCO World Heritage List, although they have been on Indonesia's Tentative List since 2009. Agnew et al, reporting on the Getty programs, emphasized this need to more fully and comprehensively document rock art sites.[43] Australia's 2016 *State of the Environment Report* found that 'Australia's Indigenous heritage remains inadequately documented and protected, and incremental destruction continues' and that 'the systems we use to manage our heritage continue to be cumbersome: land reserves, inventories and statutes'.[44] This situation is reflected internationally. For instance, the Canadian government's Standing Committee on Environment and Sustainable Development found that 'current solutions for protecting heritage places are generally not suited to preserving Indigenous heritage places' and that they are 'poorly suited to the context of Indigenous peoples'.[45] Similarly, in the United States, legal acts to protect Indigenous sites are patchwork, lack enforcement mechanisms and 'have proven insufficient to protect sacred sites', particularly those located on federal land.[46]

Preminghana Indigenous Protected Area (IPA), a 540-hectare property in Australia's far northwest Tasmania, is recognised for its many sites of natural and cultural significance, including world-recognised petroglyphs, and artefact scatters. The risks to the area from climate change-related sea level rise are rated as high to very high. Should sea level rises occur, much of the landscape, and many cultural sites would be inaccessible, making it difficult, if not impossible to practice traditional activities. This would in turn threaten *tunapri*, which is the knowledge 'that provides for a holistic understanding of the country and peoples connection to it'. This knowledge is derived from

> understanding traditional activities and places, important elements of country that require care, the pursuit of further knowledge of ancestral use of the land and the continuation and sharing of existing knowledge held within the Aboriginal community.[47]

Increased wildfire activity and severity

Rising sea levels are not the only concern for rock art. Impacts from global warming that are creating drier and hotter environments are increasing the extent of recent wildfire activity across the globe.

In Australia, rock art sites were casualties in the 2019–2020 (Black Summer) bushfires which burned across millions of hectares. In some instances, damage was less than expected. In the Yengo National Park in New South Wales, the survey team came across rock art sites, including a cave shelter with hand stencils and engravings of circles and animal tracks that are thought to be more than 7,000 years old. For Aboriginal elder Warren Taggart, who led the survey, 'There is nothing better than finding an unrecorded site where no person has been there for thousands of years'.[48] In the nearby Wollemi National Park, where a mega-fire burnt through more than 500,000 hectares, most of the caves and engravings were protected as a result of their position relative to the winds that were driving the fire's direction, but one cave with handprint stencils was 'completely burnt out, with only one print remaining'.[49]

In other parts of Australia, thousands of ancient Aboriginal sites were adversely affected by the raging fires and damaged by the heat and soot that emanated from them. Examinations confirmed these fears. In Anaiwan country in northern New South Wales, Traditional Owners and archaeological rock art researchers found large

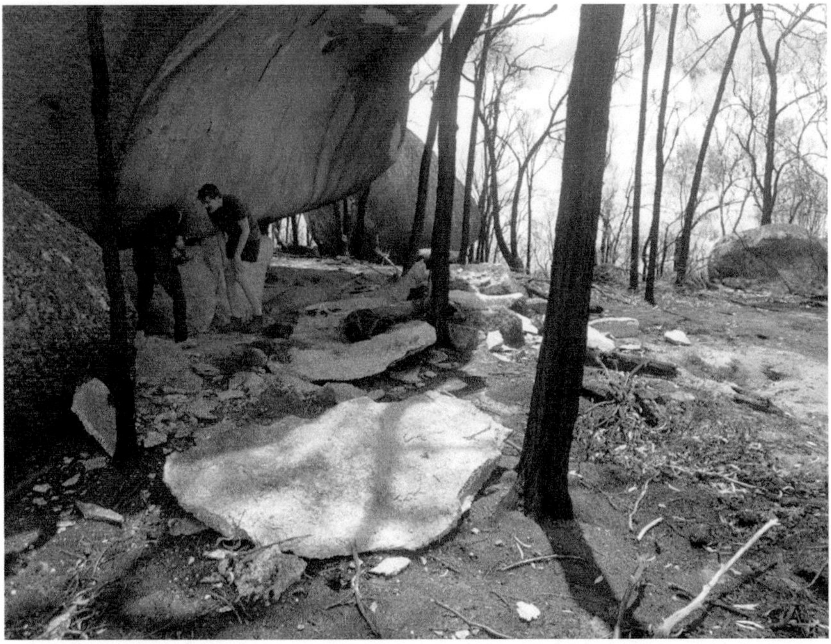

FIGURE 4.2 Anaiwan Elder Steven Ahoy and Anaiwan Traditional Owner Callum Clayton-Dixon inspect fire damage to a rock art site on the New England Tablelands. Photo by Mark Moore

granite boulders split by the heat, with artwork spalling off in large sheets and smaller fragments detaching from the surfaces (Fig. 4.2).[50] In Carnavon Gorge in Queensland in late 2018, a fire caused the recycled plastic material used in the boardwalks and viewing platforms surrounding the caves to ignite and then explode. A similar disaster had occurred a decade earlier, in 2008, at the Nganalang rock art site, in the Keep River National Park, in the Northern Territory with a recycled plastic boardwalk. At both sites, the recycled plastic was claimed to have been treated with fire retardant, but heat from the fire reached a level that rendered retardant inadequate.[51] Across Australia, painted rock art and petroglyph art sites have been damaged by bushfires, hazard reduction control burns, campfires and intentional vandalism by fire.[52] Replacing flammable material with non-flammable material for boardwalks and viewing platforms reduces combustible material near the sites. Managing regrowth and ensuring hazard reduction of undergrowth are also sensible and important tools. These are all straightforward solutions to this aspect of fire risk to rock art sites.

The problem climate change presents however is much more insidious. Increases in storm-related weather events increase the risk of lightning-strike-induced bushfires. While fires may not affect rock art sites directly, the heat dries the soil, destroying the top 25 mm of organic material that helps to bind the soil. The reduction in overhead foliage and the loss of binding material in the soil increases the opportunity for soil to be either dried out and blowing away as dust, or being washed away in the next rains.[53] Droughts have a similar effect. Activities such as fracking of artesian wells compound this.

Microorganisms and algae

Biodeterioration of rock art is a known cause of damage to painted and engraved images. The diversity of lichens and microorganisms means that some will flourish with rising temperatures that result from climate change, while others will disappear. A range of treatment approaches have been used over the years with varying success Their physiological structure means they are unable to regulate their water content, and as a result are heavily dependent upon the surrounding environment. This makes them 'among the most sensitive organisms to environmental change'.[54] Rubio-Salcedo et al's study of projected impacts of climate change on lichens in southern Europe, showed that over time vulnerable species are likely to lose more than 70% of their preferred environment, potentially leading 'to global extinction' not only because of a changed climate but also associated factors including pollution and urbanisation.[55] As Tratebas points out, 'The issue for rock art is not so much whether the lichen will deteriorate the rock art as (1) whether under natural conditions the lichen is expanding or receding, (2) if expanding, how fast, and (3) how damaging the specific lichen is to the surface'.[56] She adds a second issue that needs to be considered which 'is whether removal of lichens would cause the rock art to deteriorate more rapidly'. With such a complex mix of factors to consider, Tratebas encourages 'the input of a lichenologist to fully identify and assess the specific site situation, including viability of the species involved and knowledge of their growth patterns and preferred habitat'. She also recommends that due consideration be given to the data 'lichens themselves may contain' before their removal is attempted. Considering these and other factors including the range of ethical considerations relevant to any conservation or site management approach, she concludes that 'Rock art sites exist in a state of nature, and it would be better to leave them there, than to apply untested treatments that result in unacceptable results'.[57]

Principles and practices for rock art preservation

Carmichael et al reported that:

> Hundreds of thousands of significant archaeological and cultural heritage sites (cultural sites) along the coasts of every continent are threatened by seal level rise, and many will be destroyed.[58]

Unlike objects in museums and heritage buildings, rock art cannot be battened down during a cyclone, or pumped out during a flood. As Paul Taçon argues 'unlike built museums and galleries, the ability to continue to care for many of these important places is limited' as there is simply not the 'high security, protection from the elements or lots of financial support like built art galleries'. He notes that: 'Current conservation methods fail because of a lack of a coordinated approach, lack of cultural context, reactionary methods, lack of training, minimal research and mismanagement'.[59] The Getty rock art program identified eight principles and six key actions that were considered most impactful in the preservation of rock art. Central to these is shared, participatory decision-making, directly informed by the views and wishes of the Traditional Owners of the site, and implemented through an agreed series of actions and documentation.

- Principle 1 – Work actively to promote rock art as a valuable heritage for everyone, and allocate sufficient resources specifically to its future care;
- Principle 2 – Manage to protect all values;
- Principle 3 – Preserve and manage rock art as an inherent part of the landscape;
- Principle 4 – Safeguard cultural rights and practices;
- Principle 5 – Involve and empower Indigenous owners and local communities in decisions about rock art management and conservation;
- Principle 6 – Use recognized ethics, protocols and standards for documentation, conservation and interpretation as the basis for management practice;
- Principle 7 – Give priority to preventive and protective conservation;
- Principle 8 – Make effective communication and collaboration a central part of management.[60]

These principles align with the role of ICOMOS' International Committee on Rock Art (CAR) in initiating and co-ordinating 'world-wide actions' to improve 'the knowledge, the protection and rehabilitation' of rock art.[61] CAR identifies its role as specifically 'raising the awareness of States to this heritage':

- By promoting international co-operation in this field
- By establishing links between researchers and specialist
- By providing advice to international organisations; by promoting and producing publication
- By creating a world inventory of rock art
- By developing traditional methods and modern digital techniques for documentation and preservation of rock art, including those which are inexpensive, non-tactile and non-intrusive, as well as protective devices.

When, in 2015, Agnew et al reported to the Getty Conservation Institute on the six actions that should be taken to preserve rock art, the need for a response to the impacts of climate change was not one of them. When considered with the Getty's eight principles and applied in the current climate crisis context these actions remain relevant.[62]

1. Improved access to information and advice

It remains the case that there is insufficient information available about the threats to rock art sites from climate change, so in addition to improving access to information and advice, a program to collect and include information about the impacts of climate change on specific sites would be a useful extension of current information and advice.

2. Community outreach and awareness programs

Local communities know how their climate is changing, so community outreach and awareness programs could include much better two-way documentation of the condition of, and threat to, rock art sites. Expanding current programs to include climate change and ensuring even coverage of programs across communities requires coordination that is not currently in place.

3. Demonstrating best practice

This seems straightforward. Best practice in the documentation, examination, study, analysis, care and maintenance of rock art sites should incorporate considerations of climate change at every step.

Best practice incorporates Indigenous knowledge and two-way learning opportunities in each aspect of research and proposed responses to the threat to rock art from climate crises.

4. Coordinated training curriculum

As Carmichael et al identify 'Many communities … are not given the opportunity to participate in a cultural site adaptation program … because no program exists'.[63] As Chapter 8 in this book identifies, climate change is not yet embedded in conservation curricula, although there is a trajectory that indicates an increasing interest in doing so. Aligning conservation curricula and pedagogies so that broader responses of conservation to the challenges of climate change can be identified is an important step. Curriculum and training is best achieved through active two-way learning partnerships.

5. Effective databases for rock art site information

Databases on rock art contain scientific data to support exercises in dating, materials analysis and other scientific endeavour. Ensuring standards are in place to record data that is relevant to climate change and rock art requires a global approach, of the kind suggested by the Getty. Engendering a global approach that supports local knowledge and agency is urgent.

6. Support for local community rock art projects

Many rock art sites attract tourism. While local communities have much to gain from tourism, much is at stake if this is not properly managed. Tourism provides a good opportunity for two-way education. Bringing interested external parties onsite enables them to better understand the threats to rock art and need for its preservation. It is also an effective tool in engaging public support.

Rock art is a significant heritage resource, yet only a very small proportion has been identified, and even less has been properly studied. Climate change represents a major risk to rock art. We place guards in the repositories that hold paintings on canvas and panels and drawings on paper and parchment, and protect their environments with a range of technical solutions. How best to guard the much older and more encyclopaedic art painted and etched on rock is a major challenge that requires local, national and international efforts. This chapter has drawn on international work being done to understand loss and responses to the loss of rock art. In doing so, it foregrounds the social contexts of the impacts of climate change on cultural heritage and highlights the central importance of understanding the cultural value of rock art before deciding on actions. Finally, despite work done to date, it is clear there remains the need for a coordinated response to climate change-induced risks to rock art.

Notes

1 Brumm, A, Oktaviana, AA, Burhan, B, Hakim, B, Lebe, R, Zhao, J, Sulistyarto, PH, Ririmasse, M, Adhityatama, S, Sumantri, I & Aubert, M 2021, 'Oldest cave art found in Sulawesi', *Science Advances,* Vol. 7, No. 3, pp. 1–12.
2 Gunn, RG, Douglas, LC & Whear, RL 2011, 'What bird is that? Identifying a Probable Painting of *Genyornis newtoni* in Western Arnhem Land', *Australian Archaeology,* Vol. 73, No. 1, pp. 1–12.
3 Cobden, R, Clarkson, C, Price, GJ, David, B, Geneste, J-M, Delannoy, J-J, Barker, B, Lamb, L & Gunn, RG 2017, 'The identification of extinct megafauna in rock art using geometric morphometrics: a *Genyornis newtoni* painting in Arnhem Land, northern Australia?', *Journal of Archaeological Science,* vol 87, p. 98.
4 Murray, P & Chaloupka, G 1984, 'The Dreamtime animals: extinct megafauna in Arnhem Land rock art', *Archaeology in Oceania,* Vol. 19, No. 3, pp. 105–16
5 Helgen, KM, Wells, RT, Kear, BP, Gerdtz, WR & Flannery, TF 2006, 'Ecological and evolutionary significance of sizes of giant extinct kangaroos', *Australian Journal of Zoology,* Vol. 54, No. 4, p. 295.
6 Murray & Chaloupka 1984.
7 Brandi, EJ 1972, 'Thylacine designs in Arnhem Land rock paintings', *Archaeology and Physical Anthropology in Oceania,* Vol. 7, No. 1, pp. 24–30.
8 Taçon, P & Webb, S 2017, 'Art and megafauna in the Top End of the Northern Territory, Australia: illusion or reality?' in Bruno David, Paul SC Taçon, Jean-Jacques Delannoy & Jean-Michel Geneste (eds) *The Archaeology of Rock Art in Western Arnhem Land, Australia,* ANU Press, Canberra, p. 145.
9 Flinders 1814 (see Estensen 2002), Warner 1937, Macknight 1969, Keen 2003, Langton 2011, Clark and May 2013 to name only a few.

10 Taçon, P & May, SK 2013, 'Rock art evidence for Macassan– Aboriginal contact in north-western Arnhem Land' in M Clark & Sally K May (eds) *Macassan History and Heritage: Journeys, Encounters and Influences*, Australian National University Press, Canberra, pp. 127–9.
11 Taçon & May 2013, p. 136.
12 Taçon, P 2018, 'Preface' in Leigh Marymor 2018, 'Australian rock art bibliography extracted from the rock art studies bibliographic database for the years 1841–2018 – part 1', *Rock Art Research,* Vol. 35, No. 2, p. 188.
13 Hayward, JA 2016, 'A mutual correspondence? Comparing the rock art of west Arnhem Land with the South Australian Museum's Australian Aboriginal material culture collection', *Journal of the Anthropological Society of South Australia*, Special Edition, Vol. 40, p. 2.
14 Hayward 2016, p. 2.
15 Hayward 2016, p. 11 citing Chaloupka, G 1993, *Journey in Time: The World's Longest Continuing Art Tradition*, Reed, Chatswood, NSW, p. 185.
16 Hayward 2016, p. 11.
17 Hayward 2016, p. 15.
18 Hayward 2016, p. 11–12.
19 Hayward 2016, p. 20.
20 Hayward 2016, p. 19–20
21 Miller, E 2021, 'Finding Fibre in the rock art and ethnography of Western Arnhem Land, Northern Australia', *Rock Art Research*, Vol. 38, No. 1, pp. 48–58.
22 Cole, N & Watchman, A 1992, 'Painting with plants: investigating fibres in Aboriginal rock paintings at Laura, north Queensland', *Rock Art Research,* Vol. 9, No. 1, pp. 27–36.
23 Cole & Watchman 1992, p. 33.
24 Veth, P, Myers, C, Heaney, P & Ouzman, S 2018, 'Plants before farming: the deep history of plant-use and representation in the rock art of Australia's Kimberley region', *Quaternary International*, Vol. 489, p. 26.
25 Huntley, J, Aubert, M, Oktaviana, AA, Lebe, R, Hakim, B, Burnhan, B, Muhammad Aksa, L, Made Geria, I, Ramli, M, Siagian, L, Brand, HEA & Brumm, A 2021, 'The effects of climate change on the Pleistocene rock art of Sulawesi', *Scientific Reports*, Vol. 11, p. 1.
26 Huntley et al 2021, p. 3 and 7.
27 Huntley et al 2021, p. 4.
28 Carmichael, B, Wilson, G, Namarnyilk, I, Nadji, S, Cahill, J, Brockwell, S, Webb, B, Bird, D & Daly, C 2020, 'A Methodology for the Assessment of Climate Change Adaptation Options for Cultural Heritage Sites', *Climate*, Vol. 8, No. 8, pp. 1–19.
29 Carmichael B 2015, 'Supporting Indigenous rangers' management of climate-change impacts on heritage sites: developing an effective planning tool and assessing its value', *The Rangeland Journal*, Vol. 37, No. 6, p. 604.
30 Carmichael 2015, p. 597.
31 Carmichael 2015, p. 604.
32 Carmichael 2015, p. 605.
33 Carmichael 2015, p. 605.
34 Giesen, MJ, Ung, A, Warke, PA, Christgen, B, Mazel, AD & Graham, DW 2014, 'Condition assessment and preservation of open-air rock art panels during environmental change', *Journal of Cultural Heritage*, Vol. 15, No. 1, p.54.
35 Giesen et al 2014, p. 52.
36 Viles, AH 2002, 'Implications of future climate change for stone deterioration', *Geological Society*, Vol. 205, pp. 407–18.
37 Giesen et al 2014, p. 52.
38 Huntley et al 2021; Taçon, PSC, May, SK, Wesley, D, Jalandoni, A, Tsang, R & Mangiru, K 2021, 'History disappearing: the rapid loss of Australian contact period rock art', *Journal of Field Archaeology*, Vol. 46, No. 2, pp. 119–31; Adger, WN, Barnett, J, Brown, K, Marshall, N & O'Brien, K 2012, 'Cultural dimensions of climate change

impacts and adaptation', *Nature Climate Change,* Vol. 3, pp. 112–7; Carmichael, B, Wilson, G, Namarnyilk, I, Nadji, S, Brockwell, S, Webb, B, Hunter, F & Bird, D 2018, 'Local and Indigenous management of climate change risks to archaeological sites', *Mitigation and Adaptation Strategies for Global Change,* Vol. 23, No. 2, pp. 231–55.

39 Cubasch, U, Wuebbles, D, Chen, D, Facchini, MC, Frame, D, Mahowald, N & Winther, JG 2013, 'Introduction' in TF Stocker, D Qin, G-K Plattner, M Tignor, SK Allen, J Boschung, A Nauels, Y Xia, V Bex & PM Midgley (eds) *Climate Change 2013: The Physical Science Basis. Contribution of Working Group I to the Fifth Assessment Report of the IPCC,* Cambridge University Press, Cambridge, UK, p. 124.

40 Cubash et al 2013, p. 125.

41 Agnew, N, Deacon, J, Hall, N, Little, T, Sullivan, S & Taçon, P 2015, *Rock Art: A Cultural Treasure at Risk,* The Getty Conservation Institute, Los Angeles, p. 2.

42 Marzeion, B & Levermann, A 2014, 'Loss of cultural world heritage and currently inhabited places to seal-level rise', *Environmental Research Letters,* Vol. 9, No. 3, pp. 1–7.

43 Agnew et al 2015.

44 Mackay, R 2016, 'Heritage' in *Australia State of the Environment 2016,* Australian Government Department of the Environment and Energy, Canberra, https://soe.environment.gov.au/theme/heritage

45 Schulte, D 2017, 'Preserving Canada's heritage: the foundation for tomorrow', *Report of the Standing Committee on Environment and Sustainable Development,* House of Commons, Canada, pp. 42–43, https://www.ourcommons.ca/Content/Committee/421/ENVI/Reports/RP9295003/envirp10/envirp10-e.pdf

46 Barclay, SH & Steele, M 2021, 'Rethinking protections for Indigenous sacred sites', *Harvard Law Review,* Vol. 134, No. 4, p. 1320.

47 Tasmanian Aboriginal Centre, 2015, *Preminghana: Healthy Country Plan 2015,* p. 16, http://tacinc.com.au/wp-content/uploads/2015/07/20150529_Preminghana-Healthy-Country-Plan-_-Final.pdf

48 Cronshaw, D 2021 'New Aboriginal art identified in Yengo and Wollemi national parks in bushfire surveys', *The Cessnock Advertiser,* 12 March, https://www.cessnockadvertiser.com.au/story/7164209/new-sacred-aboriginal-sites-found-in-the-hunter-bush/?cs=459

49 Cronshaw 2021.

50 Abrams, AR 2020, 'Experts fear Australia's raging wildfires have destroyed 500-year-old Indigenous rock art in New South Wales', *Artnet,* 3 February 2020, https://news.artnet.com/art-world/indigenous-art-sites-damaged-australian-wildfires-1768107.

51 Taçon, PSC 2019, 'The aftermath of fire damage to important rock art at the Baloon Cave tourist destination, Carnarvon Gorge, Queensland, Australia', *Bradshaw Foundation,* 24 November 2019, https://www.bradshawfoundation.com/rockartnetwork/baloon_cave.php

52 Lambert, D & Welsh, B 2011, 'Fire and rock art', *Rock Art Research,* Vol. 28, No. 1, pp. 45–48.

53 Gunn, RG 2011, 'The impact of bushfires and fuel reduction burning on the preservation of shelter rock art', *Rock Art Research,* Vol. 28, No. 1, p. 56.

54 Rubio-Salcedo, M, Psomas, A, Prieto, M, Niklaus, E, Zimmermann, NK & Martinez, I 2017 'Case study of the implications of climate change for lichen diversity and distributions', *Biodiversity Conservation,* Vol 26, p. 1122.

55 Rubio-Salcedo et al 2017, p. 1122.

56 Tratebas, AM 2004, 'Biodeterioration of prehistoric rock art and issues in site preservation' in L St. Clair & M Seaward (eds) *Biodeterioration of Stone Surfaces,* Kluwer Academic Publishers, p. 202.

57 Tratebas 2004, p. 222.

58 Carmichael et al 2018, p. 231.

59 Taçon, PSC 2014, 'Australian rock art is threatened by a lack of conservation', *The Conversation,* 17 October 2014, https://theconversation.com/australian-rock-art-is-threatened-by-a-lack-of-conservation-32900

60 Agnew et al 2015, p. 4.
61 Australia ICOMOS 2022, 'CAR – rock art', https://australia.icomos.org/get-involved/international-scientific-committees/car-rock-art/
62 Agnew et al 2015, p. 5.
63 Carmichael et al 2018, p. 233.

References

Abrams, AR 2020, 'Experts fear Australia's raging wildfires have destroyed 500-year-old Indigenous rock art in New South Wales', *Artnet*, 3 February 2020, https://news.artnet.com/art-world/indigenous-art-sites-damaged-australian-wildfires-1768107

Adger, WN, Barnett, J, Brown, K, Marshall, N & O'Brien, K 2012, 'Cultural dimensions of climate change impacts and adaptation', *Nature Climate Change*, Vol. 3, pp. 112–117, doi:10.1038/nclimate1666

Agnew, N, Deacon, J, Hall, N, Little, T, Sullivan, S & Taçon, P 2015, *Rock Art: A Cultural Treasure at Risk*, The Getty Conservation Institute, Los Angeles.

Australia ICOMOS 2022, 'CAR – rock art', https://australia.icomos.org/get-involved/international-scientific-committees/car-rock-art/

Barclay, SH & Steele, M 2021, 'Rethinking protections for Indigenous sacred sites', *Harvard Law Review*, Vol. 134, No. 4, pp. 1295–1359.

Brandi, EJ 1972, 'Thylacine designs in Arnhem Land rock paintings', *Archaeology and Physical Anthropology in Oceania*, Vol. 7, No. 1, pp. 24–30.

Brumm, A, Oktaviana, AA, Burhan, B, Hakim, B, Lebe, R, Zhao, J, Sulistyarto, PH, Ririmasse, M, Adhityatama, S, Sumantri, I & Aubert, M 2021, 'Oldest cave art found in Sulawesi', *Science Advances*, Vol. 7, No. 3, pp. 1–12, doi:10.1126/sciadv.abd4648

Carmichael B 2015, 'Supporting Indigenous rangers' management of climate-change impacts on heritage sites: developing an effective planning tool and assessing its value', *The Rangeland Journal*, Vol. 37, No. 6, pp. 597–607, doi:10.1071/RJ15048

Carmichael, B, Wilson, G, Namarnyilk, I, Nadji, S, Brockwell, S, Webb, B, Hunter, F & Bird, D 2018, 'Local and Indigenous management of climate change risks to archaeological sites', *Mitigation and Adaptation Strategies for Global Change*, Vol. 23, No. 2, pp. 231–255, doi:10.1007/s11027-016-9734-8

Carmichael, B, Wilson, G, Namarnyilk, I, Nadji, S, Cahill, J, Brockwell, S, Webb, B, Bird, D & Daly, C 2020, 'A methodology for the assessment of climate change adaptation options for cultural heritage sites', *Climate*, Vol. 8, No. 8, pp. 1–19, doi:10.3390/cli8080088

Chaloupka, G 1993, *Journey in time: the world's longest continuing art tradition*, Reed, Chatswood, NSW.

Clark, M & May, SK (eds) 2013, *Macassan History and Heritage: Journeys, Encounters and Influences*, Australian National University Press, Canberra.

Cobden, R, Clarkson, C, Price, GJ, David, B, Geneste, J-M, Delannoy, J-J, Barker, B, Lamb, L & Gunn, RG 2017 'The identification of extinct megafauna in rock art using geometric morphometrics: a *Genyornis newtoni* painting in Arnhem Land, northern Australia?', *Journal of Archaeological Science*, Vol. 87, pp. 95–107, doi:10.1016/j.jas.2017.09.013

Cole, N & Watchman, A 1992, 'Painting with plants: investigating fibres in Aboriginal rock paintings at Laura, north Queensland', *Rock Art Research*, Vol. 9, No. 1, pp. 27–36.

Cronshaw, D 2021 'New Aboriginal art identified in Yengo and Wollemi national parks in bushfire surveys', *The Cessnock Advertiser*, 12 March, https://www.cessnockadvertiser.com.au/story/7164209/new-sacred-aboriginal-sites-found-in-the-hunter-bush/?cs=459

Cubasch, U, Wuebbles, D, Chen, D, Facchini, MC, Frame, D, Mahowald, N & Winther, JG 2013, 'Introduction' in TF Stocker, D Qin, G-K Plattner, M Tignor, SK Allen, JBoschung, A Nauels, Y Xia, V Bex & PM Midgley (eds) *Climate Change 2013: The Physical Science Basis. Contribution of Working Group I to the Fifth Assessment Report of the IPCC*, Cambridge University Press, Cambridge, UK, pp. 119–158.

Estensen, M 2002, *The Life of Matthew Flinders*, Allen & Unwin, Crows Nest, NSW.

Giesen, MJ, Ung, A, Warke, PA, Christgen, B, Mazel, AD & Graham, DW 2014, 'Condition assessment and preservation of open-air rock art panels during environmental change', *Journal of Cultural Heritage*, Vol. 15, No. 1, pp. 49–56, doi:10.1016/j.culher.2013.01.013

Gunn, RG 2011, 'The impact of bushfires and fuel reduction burning on the preservation of shelter rock art', *Rock Art Research*, Vol. 28, No. 1, pp. 53–69.

Gunn, RG, Douglas LC & Whear, RL 2011, 'What bird is that? Identifying a probable painting of *Genyornis newtoni* in Western Arnhem Land', *Australian Archaeology*, Vol. 73, No. 1, pp. 1–12, doi:10.1080/03122417.2011.11961918

Hayward, JA 2016, 'A mutual correspondence? Comparing the rock art of west Arnhem Land with the South Australian Museum's Australian Aboriginal material culture collection', *Journal of the Anthropological Society of South Australia*, Special Edition, Vol. 40, pp. 1–34.

Helgen, KM, Wells, RT, Kear, BP, Gerdtz, WR & Flannery, TF 2006, 'Ecological and evolutionary significance of sizes of giant extinct kangaroos', *Australian Journal of Zoology*, Vol. 54, No. 4, pp. 293–303, doi:10.1071/ZO05077

Huntley, J, Aubert, M, Oktaviana, AA, Lebe, R, Hakim, B, Burnhan, B, Muhammad Aksa, L, Made Geria, I, Ramli, M, Siagian, L, Brand, HEA & Brumm, A 2021, 'The effects of climate change on the Pleistocene rock art of Sulawesi', *Scientific Reports*, Vol. 11, pp. 1–10, doi:10.1038/s41598-021-87923-3

Keen, I 2003, *Aboriginal Economy and Society: Australia at the Threshold of Colonisation*, Oxford University Press, Melbourne.

Kimberley Foundation Australia, 2022, 'Rock art Australia: uncovering our history', https://rockartaustralia.org.au/about-us/

Lambert, D & Welsh, B 2011, 'Fire and rock art', *Rock Art Research*, Vol. 28, No. 1, pp. 45–48.

Langton, M 2011, *Trepang: China and the Story of the Macassan-Aboriginal Trade*, University of Melbourne, Carlton, Victoria.

Mackay, R 2016, 'Heritage' in *Australia State of the Environment 2016*, Australian Government Department of the Environment and Energy, Canberra, https://soe.environment.gov.au/theme/heritage

Macknight, CC 1969, *The Farthest Coast*, Melbourne University Press, Carlton, Victoria.

Marzeion, B & Levermann, A 2014, 'Loss of cultural world heritage and currently inhabited places to seal-level rise', *Environmental Research Letters*, Vol. 9, No. 3, pp. 1–7, doi:10.1088/1748-9326/9/3/034001

Miller, E 2021, 'Finding fibre in the rock art and ethnography of Western Arnhem land, Northern Australia', *Rock Art Research*, Vol. 38, No. 1, pp. 48–58.

Murray, P & Chaloupka, G 1984, 'The Dreamtime animals: Extinct megafauna in Arnhem Land rock art', *Archaeology in Oceania*, Vol. 19, No. 3, pp. 105–116, doi:10.1002/j.1834-4453.1984.tb00089.x

Rubio-Salcedo, M, Psomas, A, Prieto, M, Niklaus, E, Zimmermann, NK & Martinez, I 2017 'Case study of the implications of climate change for lichen diversity and distributions', *Biodiversity Conservation*, Vol 26, pp. 1121–1141, doi:https://link.springer.com/article/10.1007/s10531-016-1289-1

Schulte, D 2017, 'Preserving Canada's heritage: the foundation for tomorrow', *Report of the Standing Committee on Environment and Sustainable Development*, House of Commons, Canada, https://www.ourcommons.ca/Content/Committee/421/ENVI/Reports/RP9295003/envirp10/envirp10-e.pdf

Taçon, P & May, SK 2013, 'Rock art evidence for Macassan– Aboriginal contact in northwestern Arnhem Land' in M Clark & SK May (eds) *Macassan History and Heritage: Journeys, Encounters and Influences*, Australian National University Press, Canberra, pp. 127–140.

Taçon, P 2014, 'Australian rock art is threatened by a lack of conservation', *The Conversation*, 17 October 2014, https://theconversation.com/australian-rock-art-is-threatened-by-a-lack-of-conservation-32900

Taçon, P & Webb, S 2017, 'Art and megafauna in the Top End of the Northern Territory, Australia: illusion or reality?' in B David, PSC Taçon, J-J Delannoy & J-M Geneste (eds) *The Archaeology of Rock Art in Western Arnhem Land, Australia*, ANU Press, Canberra, pp. 145–162.

Taçon, P 2018 'Preface' in Leigh Marymor 2018, 'Australian rock art bibliography extracted from the rock art studies bibliographic database for the years 1841-2018 – part 1', *Rock Art Research*, Vol. 35, No. 2, pp. 188–248.

Taçon, P 2019, 'The aftermath of fire damage to important rock art at the Baloon Cave tourist destination, Carnarvon Gorge, Queensland, Australia', *Bradshaw Foundation*, 24 November 2019, https://www.bradshawfoundation.com/rockartnetwork/baloon_cave.php

Taçon, PSC, May, SK, Wesley, D, Jalandoni, A, Tsang, R & Mangiru, K 2021, 'History disappearing: the rapid loss of Australian contact period rock art', *Journal of Field Archaeology*, Vol. 46, No. 2, pp. 119–131.

Tasmanian Aboriginal Centre, 2015, *Preminghana: Healthy Country Plan 2015*, http://tacinc.com.au/wp-content/uploads/2015/07/20150529_Preminghana-Healthy-Country-Plan-_-Final.pdf

Tratebas, AM 2004, 'Biodeterioration of prehistoric rock art and issues in site preservation' in L St. Clair & M Seaward (eds) *Biodeterioration of Stone Surfaces*, Kluwer Academic Publishers, pp. 195–228.

Veth, P, Myers, C, Heaney, P & Ouzman, S 2018, 'Plants before farming: the deep history of plant-use and representation in the rock art of Australia's Kimberley region', *Quaternary International*, Vol. 489, pp. 26–45, doi:10.1016/j.quaint.2016.08.036

Viles, AH 2002, 'Implications of future climate change for stone deterioration', *Geological Society*, Vol. 205, pp. 407–418, doi:10.1144/GSL.SP.2002.205.01.29.

Warner, W 1937, *A Black Civilization: A Social Study of an Australian Tribe*, Harper & Brothers, London.

5

ORAL HISTORIES OF NATURAL DISASTERS IN TIMOR-LESTE

Sara Soares and Robyn Sloggett

The memories and histories that are told and retold by diasporic communities represent significant community resources. Oral histories play an important part in strengthening intergenerational cohesion in communities that have experienced substantial disruption. As a form of community heritage, they are a valuable resource in the process of nation-building, and provide a continued sense of identity in the midst of disruption and change. As this chapter demonstrates, oral histories also provide information that is relevant in addressing the challenges produced by natural disasters and climate change.

Migration resulting from climate emergencies is becoming an increasing reality for many Small Island populations. It has become an area of particular interest to governments and non-government organisations that want to understand how migrant populations manage resilience and adaptation. The impacts of migration can be devastating as people are forced away from the places and institutions where their heritage is located. Communities are fragmented as the result of forced migration and cultural and social knowledge is disrupted. This loss, and associated fragmentation, is balanced by oral traditions and histories that continue to be passed on down generations. This chapter examines the memory of natural disasters that the East Timorese diasporic community in Australia retains from their experiences in Timor-Leste. It brings together three aspects of responses to natural disasters, memory and the construction of memory, government planning, and research. This chapter also examines the ways in which inherited oral histories retain and accrue value as a form of community heritage. Many of these memories and histories relate to the environment, to natural events including disasters, and associated traditional and customary practices. In this respect, these oral histories are a form of heritage that link responses to times of crisis across generations. They reflect how resilience and adaptation is managed in the face of natural disasters, and

DOI: 10.4324/9781003163312-6

provide information on how these responses can be incorporated into planning and into the preservation of diasporic-held heritage.

Despite the social disruption experienced by East Timorese people (or perhaps because of this) individual memories of events, and of stories told to family members remain strong. In 2012, a series of interviews was conducted to capture heritage accounts of natural disasters in Timor-Leste amongst East Timorese diaspora living in Australia and Timorese in Timor-Leste. This research was part of a larger project that sought to understand the impact and occurrence of natural hazards in Timor-Leste. The authors interviewed members of the Timorese community in Melbourne and Sydney, recording their experiences of natural hazards in Timor-Leste. Interviewees shared memories and histories of floods, landslides, earthquakes and tsunamis. While earthquakes and tsunamis are not climate-related hazards, there is evidence that climate change may increase the frequency of these events and worsen their impact, and so these were included in the interviews.[1] The results of the interviews demonstrate that, in terms of response to natural disasters, populations are unlikely to differentiate climate change-induced events from other natural disasters. Rather they maintain a broader ontological framework in which to understand and locate such events.

The Democratic Republic of Timor-Leste became the first new sovereign state of the millennium, on 20 May 2002 after 24 years of Indonesian occupation and 3 years of UN transitional administration. As a result, many of the instruments of government are of very recent construction. A long history of colonization, and the dismantling of civic institutions and the destruction of infrastructure immediately after the 1999 Popular Consultation, has meant that the vast majority of archival records relating to Timor-Leste are held overseas and focus on areas of interest to the two colonial powers, Portugal and Indonesia. Many Timorese, however, can report historical knowledge through oral histories that have continued to keep history alive. Immense social disruption over the past 50 years has created a large Timorese diasporic population, which increasingly includes next generations who were not born in Timor-Leste. As the floods in Timor-Leste in 2021 have demonstrated, diasporic communities are important in supporting rebuilding of local infrastructure through donations and other forms of at-distance support.

The Indonesian occupation of Timor-Leste from 1975 was characterised by intensive Indonesian-led military control and surveillance of local populations, widespread abuses of human rights and loss of lives, and a long campaign of opposing guerrilla warfare by Timorese forces. Many East Timorese fled to Portugal and Australia. The Commission for Reception, Truth and Reconciliation (CAVR) conducted an extensive inquiry from 2001 to 2005. The resulting 2,500-page report, *Chega!*, found that at least 1 in 7 of a pre-invasion population of less than 700,000 died as a result of the war. Of the survivors many suffered from widespread and systematic crimes against humanity and war crimes over the 25-year period of occupation to 1999.[2] On 30 August 1999, a UN-sponsored referendum found almost 80% of East Timorese supported independence from Indonesia. This result sparked violent retaliation from Indonesian-backed militia with the destruction of 90% of the nation's physical infrastructure. It is estimated that more than 450,000 people were displaced by violence, including roughly 240,000 people to Indonesia, accounting for about 25% of the population.[3] Thousands fled to Australia. Almost all sectors of society were affected.[4] Timor-Leste had few trained personnel, education and health services collapsed, 80% of the country's primary schools were destroyed, and an estimated 70% of private homes and public buildings were laid waste. Valuable documents, including cultural heritage rehabilitation plans, civil registry and health and education records, were also destroyed.

Timor-Leste's traumatic recent history is significant to understanding the country's economic and hazard vulnerability during this current period of national rebuilding. Reconstruction in all areas continues and is being delivered with strong national spirit and skill, despite weak infrastructure and limited resources. However, natural disasters and climate change impacts have the potential to erode these hard-won development gains.

Natural hazard and climate change risk in Timor-Leste

Timor-Leste is one of the world's most vulnerable countries to natural disasters due to its geographic location, topography and socioeconomic conditions. The country is located near one of the most active tectonic plate boundaries in the world, at the point where the Eurasian and the Australian plates meet, and as a result experiences high seismic activity.[5] Earthquakes are common and cause significant damage, where slopes are unstable earthquakes can trigger extensive landslips with a negative impact on peoples' lives, livestock, roads, infrastructure and property.[6] Timor-Leste's proximity to the equator means it is also at risk from tropical cyclones, experiencing seven between 1969 and 2011. Its climate is strongly affected by the West Pacific Monsoon and the Southern Oscillation and 'El Niño' (ENSO), which can produce severe droughts as well as flooding.[7] In some parts of the country, this has adverse consequences for crops and subsequence food shortages.[8] Estimates suggest that an extreme drought during the 2015/2016 El Niño affected nearly half of national maize and rice production, and more than half of the affected population lost livestock because of heat impacts and water scarcity.[9]

The geography of Timor-Leste is very rugged, with a central spine of mountains running along the length of the island. The mountains descend directly to the sea on the north coast and on the south are coastal plains. Overall, the soil quality is alkaline and poor with limited nutrient and water holding capacity. Approximately 44% of the country's land mass has slopes greater than 40°. Seasonal monsoon rains falling on steep slopes cause frequent flash flooding and landslides, which regularly damage and destroy infrastructure, especially in rural areas. These events cause major disruption to the fragile road network of the country, isolating communities and even districts for days, weeks or months. The country is prone to landslides not only due to its geology, being recently uplifted and undergoing rapid erosion, but also because of inappropriate land use. Along with food prices and shortages of dietary staples, scarcity of arable land and the impact of natural hazards have been cited as the primary reasons for deterioration in food security in the country.[10]

Climate change is acknowledged to have the potential to exacerbate all weather-related hazards. Countries such as Timor-Leste where the majority of the population derives their livelihoods from subsistence agriculture activities are especially vulnerable to climate change impacts. Future climate projections predict increasingly warm air temperatures, leading to very hot days, which affect crop health; extreme rainfall is predicted, increasing the risk of flooding events, though the frequency in drought will remain unaffected; rising sea levels present a threat to coastal cities and communities; and ocean acidification is projected to increase and threaten coral reef ecosystems.[11] The sea level rise combined with natural year-to-year changes will make more noticeable the impact of storm surges and coastal flooding.[12] Heavy rainfall events of greater intensity interspersed with dry periods will also put Timor-Leste's soils at risk of greater erosion and more frequent landslides.[13] Causal links between climate change and increased risk from earthquakes and tsunamis have also been identified.[14] There is also the potential to increase the incidence of vector-borne diseases, such as malaria and dengue fever, through changes in the concentration and distribution of mosquitoes, and from more intense rainfall causing increased soil erosion, resulting in an increase in the volume of suspended sediments.[15] This, in turn, will put increased pressure on health facilities and will reduce people's ability to work and earn a livelihood.[16]

Understanding risks from natural disasters, climate change and the potential impacts on the country's development are priorities for the national government of Timor-Leste. Natural disasters have been a focal sector in most of the national climate change documents, and the government has been proactive in taking steps to improve Disaster Risk Management (DRM) and Climate Change Adaptation (CCA). This includes ratifying the United Nations Framework to Combat Climate Change (UNFCCC) and recently submitting its National Adaptation Plan (NAP) to the UNFCCC Secretariat.[17] National policies and governance structures for DRM and climate change, however, have developed separately, and the government is currently drafting new plans to better integrate the two fields. One of the core objectives of the NAP is to facilitate integration of CCA across all sectors, including DRM.

Disaster risk reduction (DRR) and CCA are critical to building the resilience of vulnerable communities, during an emergency response situation, who are disproportionately affected during periods of stress. Timor Leste is particularly vulnerable. In its recent country Climate Risk Profile (2021), the World Bank notes that climate change is a major risk to Timor-Leste's recent hard-won development gains, one that threatens to increase inequality and drive significant damage and loss.[18] A 2013 Asia Development Bank report estimated that Timor-Leste's economy would be among the worst hit in the Pacific region by climate change.

Damage resulting from tropical windstorms, landslides, floods, earthquakes and tsunamis could cost as much as 10% of annual GPD by 2100.[19] In this context, the rural poor and other marginalised groups, such as women who are already suffering from gender-based vulnerabilities, are further impacted in their ability to gain access to health and social services, and anticipate, prepare for, cope with and recover from a disaster.

Local knowledge in DRM and CCA

The importance of integrating local knowledge and practices into DRM, including CCA, is acknowledged by research and development organisations and specialists and supported by a growing body of literature.[20] The Intergovernmental Panel on Climate Change (IPCC) acknowledged Indigenous and local knowledge in its Sixth Assessment Report as a key enabler in developing climate adaptation strategies in Small Island nations such as Timor-Leste.[21] The 2004 Indian Ocean earthquake and tsunami has been credited with sparking interest in this field, however, in practice there exists little documentation of its application. Case studies from Indonesia, Philippines and Timor-Leste have demonstrated that heritage practices and knowledge are effective contributors in supporting better outcomes for communities affected by the climate crisis. They show that Indigenous and local knowledge and traditional practices can support environmental awareness, strengthen social cohesion and help communities better respond to climate impacts.[22] Mercer et al have developed a framework for integrating local knowledge to reduce a community's vulnerability to a wide range of disasters and have applied this to Timor-Leste.[23] They engaged local communities using a participatory approach to assess disaster risk, including climate-related risk. They argue that this approach could help bridge the data gap challenge, thereby supporting relevant policies and strategies for DRR and CCA and empowering East Timorese as part of the creation of the state and contributors towards the development of East Timor's national identity.

Timor-Leste's recent NAP prioritises community-level integration of DRR and CCA, and has a high consideration for traditional laws and customs when considering the baseline conditions that shape risk, vulnerability and adaptive capacity.[24] Tara bandu, traditional customary land management practices, was identified as an important entry point for engaging communities to raise awareness of government policy at the local level, and as an important mechanism for facilitating community-led CCA activities. Since the restoration of independence, tara bandu (and local language variants) has been taken up in a number of villages to protect and conserve natural resources and, as a cultural institution, it is one that the national government is keen to support. As noted by Demetrio do Amaral de Carvalho, Secretary of State for Environment: 'Tara bandu uses local knowledge of conservation and maximises the social capital of villagers … It strengthens our community bonds, and also helps to protect the environment'.[25]

Researching memories and oral traditions relating to natural hazards in Timor-Leste

Although Timor-Leste is vulnerable to a number of natural hazards, a key challenge for effective DRR and CCA is the lack of credible datasets in the country regarding disasters and impacts.[26] In 2010, the World Bank noted that to assess climatic variability and change, long-term records were needed that provided representative geographical and statistical sets of data for rainfall and temperature.[27] The records that are available were collected by colonial government organisations, created by missionaries belonging to the Catholic Church, and some anthropological records. As a result, very few records are available that could provide the data necessary to understand the frequency, magnitude and impacts of large natural events. Where they exist they reflect specific geographical interests, and are not focused on the responses of individuals which is an important part of an effective natural hazards programme.

This chapter examines the information provided by East Timorese diaspora in Sydney and Melbourne to the authors in 2012. It is also supported by historical records archived by Kevin Sherlock, an Australian who dedicated half his life to finding and collecting materials about Timor through self-funded trips across Australia, Portugal and Macau. By the time he passed away in 2014, he had amassed an invaluable collection of information on Timor-Leste. In 2012, he generously and carefully compiled a set of materials related to disasters, in the hope that his work would contribute towards the 'greater knowledge of the country and would be of use to the East Timorese in preserving and learning from the past and in planning for and working towards the building of their nation'.[28]

East Timorese Diaspora in Australia

Timor-Leste is one of Australia's closest neighbours. Prior to the Indonesian invasion in 1975 Australians regularly visited Timor-Leste (then known as Portuguese Timor) mainly for holidays, and small communities of East Timorese lived in Australia. In the period of Indonesian rule, 1975–1999, it is estimated that about 10,000 East Timorese fled to Australia, and another 10,000 to Portugal. Within this period, however, there were waves of increased migration, including between 1980 and 1986, and 1991 and 1993, in which many East Timorese moved from Portugal to Australia for economic, geographic and family reunion reasons.[29] In 1991, the infamous Santa Cruz massacre in which at least 250 mainly young East Timorese were killed during a funeral for two independence activists, led to another increase in migration to Australia.

Prior to 1975, East Timorese diaspora were mainly from the elite and well-connected families who migrated for purposes of trade or education. The Indonesian invasion caused mass migration of people across Timor-Leste, creating a greater socioeconomic mix within the diasporic communities. After 1999, education and trade again became a significant part of the diasporic story in Australia as partnerships were put in place for trade, political engagement and education.

Hajek and Goglia note the 'strong transnational link between the immigrant communities in different countries and the motherland',[30] which is in part due to the nature of East Timorese identity in the period 1975–1999, when maintaining a sense of collective identity that was both internally cohesive and publicly visible was important to progress the case for independence from Indonesia. East Timorese diaspora remained closely engaged with the struggles of their homeland during the anti-Indonesian resistance period; they raised funds and provided a base for the Diplomatic Front (Frente Diplomatica), which was eventually vital in securing the nation's restoration of independence. This strong collective identity and pride in East Timorese culture is maintained both within Timor-Leste and by East Timorese diaspora and is identified in the government's Strategic Plan as critical to Timor-Leste's future.

> We recognise that if we neglect our cultural and historic roots, and think only of the present and the material aspects of life, we will be overwhelmed by the forces of globalisation and risk losing our unique cultural identity – the very thing we fought so long and hard to preserve.[31]

Diasporas can also play a vital role in responding to a humanitarian crisis, often as part of the first response in the aftermath of a disaster due to their understanding of their country of origin and heritage, and close local ties.[32] In early April 2021, several days of heavy rainfall brought by tropical cyclone Seroja resulted in catastrophic flash floods and landslides that adversely impacted more than 30,000 households, including more than 50% of Dili households, and destroyed critical infrastructure, such as roads and bridges.[33] Maize and rice production were severely affected, with agricultural land washed away and estimates for rice crop losses as high as 60% of the total yield.[34] The scale of the floods and the ability to access immediate information through social media both within Timor-Leste and overseas prompted diasporic organizations to act quickly. Their response, sending funds raised via online fundraising campaigns and in-kind donations through informal networks and organisations to those affected, reflected increased digital communication and connection between the diaspora and young Timorese in Timor-Leste.

DEMAC (Diaspora Emergency Action & Coordination) reviewed the humanitarian response to the flooding. One of the key recommendations was to explore the potential for further coordination between the Timorese diaspora, the Timorese government and international humanitarian agencies.[35] DEMAC found the lack of coordination the diaspora had with the government and humanitarian responses was a key challenge; making donations difficult, and raising concerns about duplication of relief efforts or that goods or funds could not be delivered to appropriate recipients. DEMAC recommended that the diaspora organisations should be engaged in a way that recognises and utilises their unique strengths.

DEMAC also suggested that humanitarian agencies should invite interested diasporic organisations to future sector coordination meetings, and involve them in the recovery phase, needs assessments and after action reviews. The diaspora themselves recommended the creation of a forum of Timorese diasporic organisations, which could both improve intra-diasporic coordination and bolster their representation in future humanitarian responses.

Interviews with East Timorese Diaspora in Australia

The first set of interviews was conducted in Melbourne at the Noble Park Community Centre on 7 July 2012, hosted by the Timorese Association in Victoria, and the second in Sydney at St Elias church in the suburb of Ashcroft, and a private home in Sydney on 15 July 2012. Interviewees were identified by existing community contacts and the strong networks the research team had in the East Timorese community in Australia and in Timor-Leste. A sheet in English, Tetun and Portuguese describing the project was circulated within the community seeking involvement from those who are interested in contributing to the project. Interviews were recorded and, in accordance with Australian government ethics requirements for research, participants were given anonymity if desired (although no participant sought anonymity), could withdraw consent at any time (none did so). All participants signed a consent form to allow the information they provided to be made public. In regard to this last point, a number of participants indicated that they were taking part in the interviews because they felt the project was important and that it was vital to share knowledge that could contribute to Timor-Leste's future.

Floods

The earliest account of flooding identified during the research was in *O Porvir*, a Portuguese language newspaper printed in Hong Kong, held in Kevin Sherlock's archives. The newspaper reported a cyclone on 26 and 27 April 1898 that swept through the island causing significant damage to homes, public infrastructure and coffee plantations which were the central export crop for the local economy, and resulting in at least 60 deaths.[36] *O Porvir* reported this as the first time a cyclone of this magnitude was seen by locals on the island. The account commented on the impact the disaster would also have on Macau's economy, the nearest Portuguese colonial outpost that was often called to balance the Portuguese economy and provide disaster relief to the colonies.[37]

In the oral histories, the 1938 floods were a major focus. The first person interviewed was 81-year-old Senhora Ines Varudo. The interview was conducted in Tetun. Sra Varudo's daughter Lurdes Soares provided translation. Sra Varudo lived in Dili in the suco (village) of Santa Cruz. She recalled several flood events, including her particularly vivid memory of an event in 1938 when it rained for a week. As a 7-year-old, Sra Varudo recalled how the water came down 'from the mountains but instead of going to the ocean, the ocean also came up on shore so the whole of Dili was flooded'. The resulting floods were so severe that the only way to move around Dili at that time was by dinghy.

Sra Cecilia Sequeria was born in Baucau and moved to the Taibessi area in Dili when she was seven years of age, and left Timor Leste in 1975. She remembers being told about the 1938 flood and the damage that ensued. Senhor João Rosa, who was born in 1939, also identified the 1938 floods from accounts he was told by his grandmother, noting they occurred on 11 February 1938 and resulted from seven nights and seven days of rain. He also described the landslides that resulted 'so half the mountain slides' and destroyed the first Comoro bridge built by the Portuguese. Sr Rosa's grandmother had described what the sea and river looked like, and how there was 'too much wind, people could not communicate because of the rain'. Sr Rosa recalled being told that the flooding affected 'all of Timor'.

Sra Lurdes Soares, who left Timor-Leste in 1975, remembered another major flood that occurred in Dili in 1971 or 1972 after a week of heavy rain and floods. Her most significant memory was that

> despite all the rain, we didn't have any water, the taps were all dry. We had to go to a fountain to cart water from the mountains. The whole of Dili ran out of water ... We had to walk miles and miles up the mountains to get any fresh water, up towards Dare. It was a week of non-stop rain and the Comoro River was so high that the bridge actually fell. We didn't go to school for the whole week. But we also have a river in our neighbourhood, and I remember every day everyone would go out standing on the street to see how high the river was filling up, because we thought we were all in danger of being washed away. We didn't but we know the houses further up to the Old Hospital, more of the houses were washed away.

Earthquakes and tsunamis

Timor-Leste's location means it is vulnerable to earthquakes and tsunamis. Earthquakes, though not the most frequently occurring natural hazard, cost the country more than any other singular natural hazard.[38] German geologist, Arthur Wichmann, spent much of his career studying the geology of Indonesia. He compiled an earthquake and tsunami catalogue from 1538 to 1877 using a combination of data sources, including 350 years of observations of earthquakes and tsunamis many from Dutch colonial records. This was translated into English in 2017.[39] One notable earthquake occurred along the north coast of Timor-Leste on 13 May 1857, causing a tsunami consisting of four waves, the highest of which was 3.14 m. The tsunami inundated the north coast around Dili, and completely flooded the villages of Liquiçá and Hera. The quake was 'very strong' in Hera, Laclo, Lautem, Lale and Batugade. A mud volcano erupted near Viqueque and a hill sank on Pulau Kambing (now Atauro Island) killing more than 36 people. The quake was also felt in Kupang (West Timor) and Ambon.

Sra Ines Varudo remembered earthquakes occurring at least once every year although these were not big enough to cause damage. She recalls one where 'people were so scared that they thought it was the end of the world'. Her daughter Sra Lurdes Soares remembers one in 1973/1974, recalling the tradition of responding to earthquakes by running outside and banging pots and pans to make a loud noise.

> It was in the morning, we were just getting ready to go to school, we all rushed outside banging on everything because God must have been asleep. We have earthquakes when God, who is holding the earth, falls asleep. We needed to make enough noise to wake him up again and steady himself.

This legend is known throughout the island and was observed by a Portuguese officer on the corvette Sá da Bandeira visiting in 1870.[40] In his account, God was testing the locals by pulling on the string connected to the earth. Unless enough noise was made to prove that humanity was indeed still alive, God would destroy the world.

From a very young age, Sra Soares remembers how everyone ran outside during tremors and earthquakes: 'we couldn't stay inside because the danger of something falling'. She described the earthquakes as starting

> very small – like someone is shaking you out of bed and then gradually it feels like the whole house is going to be ripped out of the earth. Everyone gets really scared … Everyone floods out to the street and you bang on anything you can find like plates or glasses or anything that can make noise, and it goes on for a few minutes. It was the worst few minutes in my life I can tell you that. The last one I remember was just before the invasion, whether it was 74 or 75, and it was just a normal occurrence, like you have once or twice a year. The whole of Dili went through it. Everyone was affected and talked about it. I don't know if people were hurt or not but some houses got cracks. I can't remember if Mum and Dad had to do anything to fix our house, but I remember the shelves and plates moved because all the chocolates they hid fell out.

Sra Soares also described it as

> fun too because everyone was out on the streets. But it also felt like the end of times, like when you read about Noah's Ark. In the kitchen things would start falling off the shelves, and it's horrible because you can't stand still. You don't know if it is going to swing one way or the other. That was terrifying. You know you have to stand out in the clear to be out of danger of being squashed by something. By the time we hit the street we knew that if we were going then at least we would all go together. It's not something you look forward to but as soon as the earth starts to shake you know exactly what you have to do. I experienced this a few times, probably 3, before I left

> Timor-Leste when I was 12 or 13. Luckily, we always had adults with us at home, and there was always someone who would tell you when you needed to get out of the house. You would feel the distinct noise and the rattle and know exactly when you have to get out of the house. It is so entrenched in my memory. I remember for the last one I didn't have to wait for anyone to tell me to go outside, I was the first one to run.

The occurrence of a large earthquake in 1972 was confirmed by Sra Olinda Maia who lived in Balibo Villa, Bobonaro District.

> It was so strong. It felt like it could be an 8.5 or 8.6 magnitude earthquake. We were so scared that we all slept outside so our homes wouldn't fall on us. The earthquake tremors lasted from the afternoon until evening. We all ran outside and prayed to God for the earthquakes to stop. We also banged on pots, our homes, everything to make the earthquake stop.

Sra Cecilia Sequeira who lived in Baucau then in Taibesi, Dili, recounted how the traditional palm tree houses moved and recalled their stability during earthquakes.

> Normally when the earthquake begins, the dogs outside would start barking. The people would start banging and making noises with tins and everything. That's what I experienced you know, sometimes you feel like you're moving but then it passes. I think in Dili we used to have them every year. Sometimes more stronger and sometimes just lightly, that's what I remember … My house [was not damaged] because it was made from palm trees [palmeira] – they made the walls of the house from the stalks and the roof from the palm leaf. You could hear and feel the house moving but it didn't go down. I was maybe … 16 or 17 at the time. That was the strongest one I can remember. It was even longer I think in the morning. We just could feel the house just swaying. Mum and Dad used to tell us, 'don't go outside', no they didn't like us to go outside [so] we stayed under the door frame … In my time, people normally don't run … now is all brick houses, but in my time was wood. Wood is more secure. [You] could feel [the palm tree houses] swaying, nothing go up and down, but sideways. We didn't see the earth [move], we just saw the walls of the houses going like this … you could hear the whole house creaking.

In their 2016 report, the CFE-DMHA noted that there are few earthquake-resistant structures in Dili or district capitals, with the majority of private dwellings constructed as non-engineered concrete or masonry buildings. These buildings will likely have a high rate of failure in a strong earthquake, and devastating levels in a major earthquake with an epicentre close to Dili or other district capitals.[41]

Sr João Rosa recalled a large earthquake in 1997 that opened up the ground and buried coconut trees so that only the leaves on top were visible. 'If you go to the

Comoro River you can still find the coconut trees there'. Sr Rosa recognised that 'In Timor land is never stable'. He described Same being destroyed by landslides and the bridge in Comoro, which was first destroyed in 1938, also collapsing. Driving into Dili during one particularly large earthquake, Sr Rosa recalled seeing trees moving but the car being unaffected. He recalled people running out from their houses calling out 'God we still exist!'. He recounted:

> There was a big earthquake, before I married, in Dili but I think [it affected] all the country. It came with a sound like a volcano was going to erupt or something, because all the island was shaking. We ran outside because [it was] explained to us the house can collapse but when we went back inside there was no damage to the house.

Senhor Miguel Sandanha Vidal, who lived in Dili, recalled a story he had been told by his uncle, who was a fisherman in Lecidere. According to Sr Vidal, one day while his uncle was out at sea,

> suddenly he saw the water was dried up on the seas – half of the water went up and went down … and he was surprised wondering 'what happened' but then the water came back … it was in the 1970s.

He also recounted events he had heard second hand taking place in Caicoli and Comoro.

> They said the houses in Caicoli were swallowed up by the earthquake. I think in the 1930s or 40s. I'm not sure if the story is true or not but they said when they tried to rebuild the houses they kept finding pieces of buildings under the water. I think that part in Caicoli used to be full of water and now they have started building houses again. I also heard a story in Comoro that a tsunami took some places, though I can't tell you exactly which year.

Sr Rosa was sure that there had never been a tsunami in Timor-Leste 'maybe in the future but since I grew up there I've never heard about tsunamis'. During the 2004 Indian Ocean earthquake and tsunami, he was in the country and recalled there were concerns that it would reach Timor. Someone announced that the tsunami was coming, resulting in a pregnant woman running in fear and many fleeing to the mountains. However, he noted:

> The people were lying to each other, saying that the tsunami was coming, only to steal things. It was a big mess but the people living in the city and the UN, they didn't see anything.

While most people focused on their experience of floods and tsunamis, Senhora Olga Xavier Gago who comes from the village of Com in the district of Los Palos brought a much wider lens to the discussion. She described how heritage buildings provided evidence of the adaptation made by leaders in the Los Palos district to tsunamis, giving a detailed description of governance, decision-making, and leadership in Los Palos. She began 'it was interesting to me to ask those questions to my *avos*, my grandparents because that's how we grow from the tradition'. Sra Gago discussed the name for the village Tutuala:

> Tutuala originally comes from the Fataluku, Tutu hai hala. The history of that word is linked to a big tsunami that happened in that area which affected the whole village. A lot of people died during that period but I don't know how many.

The account of a tsunami in Tutuala was repeated by interviewees in Tutuala a month after Sra Gago's interview in Sydney, and identified as occurring during the World War II Japanese occupation of Timor.

Sra Gago described how traditional houses were built in accordance with strict rules, using four wooden poles to keep the structure above the ground so water can pass below. This construction ensured people's safety during floods and tsunamis 'because they were very experienced with that'. Reflecting on the oral history associated with this building design Sra Gago noted that this design was developed

> only a few hundred years ago, I don't know how many years, but because of the tsunami, the avos (grandparents, ancestors) at that time came up with the idea of how they can survive by building that type of house. You can't build a house on top of that wet area so build it on top of four poles so if another tsunami comes they can survive because they are on a higher level. The lian na'in (lord of the words), all the men, they had a big meeting. At that time my avo, my great, great grandfather, was the general secretary in Lautem. His name was Amu Antonio Fernandez Xavier. He was a leader in the village and he looked after the people. He passed away when he was 113 years. I had a chance to meet him and that's how I know this story.

Sra Gago also noted another purpose served by the traditional houses.

> During the rainy season, the waters rise up, sometimes 1 or 2m high so that the crocodiles in the river would be swimming outside. Everybody would be scared. Luckily, we have these types of houses so all the people can go up and the crocodiles can't reach them.

She also reflected the way in which the design of traditional houses led to settled patterns of living, and supported the development of good governance, as people

> felt very comfortable so they continued to build that type of house until now. So that's one of the traditions that was kept alive … There was a

> meeting where they named the village and decided who would lead the village. My great grandfather said, 'OK let's vote – who can read, or who is capable to be head of district or the head of the village, who can look out for the population, how do we run our organisation and the daily routine of our lives from now on'. So they started there and then he started to have connection with the missionaries, to ask for their help in the Los Palos area. He had the chance to have a meeting with representatives of the other 13 districts in Los Palos, so from the Bacau to Los Palos area he was considered a very good administrator and secretary general.

When asked about recent tsunamis Sra Gago recounted one 'sometime in the 1990s, my cousin told me that in the Tasi Tolu area some of the coconut trees disappeared and the land level changed a lot'. This event appears to correspond to the magnitude 6.9 earthquake that occurred in the Flores Sea, in Indonesia, on 14 May 1995.[42] This resulted in land subsidence along Timor-Lestse's western coast, from Dili to Maubara. In Marinir, Dili, a wave run-up of 4 m with flooding up to 120-m inland was reported, 11 people disappeared, 19 were injured, and houses and 40 fishing boats were damaged. Damage extended to the Maliana region. *The Indonesian Observer* reported that Dili's harbour was so badly damaged it needed to close operations temporarily.[43]

Valuing oral heritage as social capital

Nakashima, Krupnik and Rubis describe memory as a form of social capital that is exchanged between and on behalf of individuals, and particularly, across generations. They argue that:

> Because kinship relations and exchange networks themselves provide a means of mitigating impacts of climate events, it is important to understand the way these networks of connection operate and the political and economic contexts under which they will be able to continue to flourish. There has been much recent anthropological research conducted on migration, transnationals, diasporas and global flows of people and goods, which is helpful for understanding the kinds of *social capital* that people might employ to cope with climate change.[44]

Cultural heritage retained as memory and oral tradition can be prompted, through formal interviews, by access to objects and other aides de memoire, by proximity to place or the memory of place, and by continued retelling. The value of social capital that is embedded in heritage is well understood, but perhaps less understood is the potential role of memory and oral histories in informing emerging government policy in areas including disaster mitigation and climate change.

A 2017 report on risk assessment and mapping in Timor-Leste identified two of the 'Gaps and challenges for disaster risk reduction related initiatives', as the 'lack of historical time-series data for risk assessment' and the 'lack of spatially distributed data sufficient to construct hazard maps at a scale appropriate for planning and risk reduction'.[45] In 2010, the World Bank wrote that

> the severe lack of data, tools, and capacity to quantify and interpret those risks is limiting the potential means for integrating the knowledge into policy, analysis, strategy, and development planning and decision making.[46]

As a form of heritage, archives provide information that is useful in informing emerging government policy in areas including disaster mitigation and climate change. Oral histories have similar value, and as the interviews in this chapter demonstrate, some of the rich data that exists within diasporic communities could provide 'historical time-series and spatially distributed data' to usefully inform planning and risk reduction. Identifying significance, enabling access, and providing education are core components of heritage preservation, conservation and museological expertise. Creating partnerships that embrace diasporic communities in these endeavours brings together heritage knowledge, to address the need for new knowledge, in ways that really can make a difference.

The IPCC's Sixth Assessment Report notes in its Regional Fact Sheet on Small Islands that:

> Though it is clear the climate of Small Islands has and will continue to change in diverse ways, constructing climate information for Small Islands is challenging due to lack of observations and high-resolution climate projections, as well as the representation and understanding of key modes of variability and their interplay with trends.[47]

As the East Timorese diaspora who contributed to the research in this chapter demonstrate heritage held in oral accounts of natural disaster events provides a rich source of information for future research and decision-making. This chapter explored some of the ways in which oral histories can provide local, longitudinal data that improves social responses to disaster and public policy development involving DRM and CCA. As climate change forces more people from their homes to become climate refugees the ways in which stories about this disaster are told will be an important part of understanding what builds resilience and adaptation. For heritage professionals, recording, documenting, preserving and conveying the value of this resource is an important and urgent task.

Acknowledgements

This research was partly funded by a University of Melbourne Interdisciplinary Seed Funding Grant.

The authors would like to thank all those who generously and selflessly shared their time, resources and memories with the aim of helping Timor-Leste manage future disasters, and conserve its history and culture.

This chapter is dedicated to Ines de Carvalho Varudo and Milene Rosa Varudo.

Notes

1 Liu C, Linde A & Sacks I 2009, 'Slow earthquakes triggered by typhoons', *Nature*, Vol. 459, pp. 833–6, https://doi.org/10.1038/nature08042; Cunneen J, 24 January 2022, '5 ways climate change increases the threat of tsunamis, from collapsing ice shelves to sea level rise', *The Conservation,* accessed on 2 April 2022, https://theconversation.com/5-ways-climate-change-increases-the-threat-of-tsunamis-from-collapsing-ice-shelves-to-sea-level-rise-175247

2 Comissão de Acolhimento, Verdade e Reconciliação (CAVR) de Timor Leste, 2005, *Chega! The CAVR Report*, accessed on 25 September 2021, http://www.cavr-timorleste.org/en/chegaReport.htm

3 Niksch, L, McHugh, L & Margesson 2001, *East Timor Situation Report*, Congressional Research Service, Library of Congress, accessed on 25 September 2021, http://www.globalsecurity.org/military/library/report/crs/RL30975.pdf

4 Robinson, G, 2006, *East Timor 1999: Crimes Against Humanity: A Report Commissioned by the United Nations Office of the High Commissioner for Human Rights (OHCHR),* Princeton University Press, accessed on 2 April 2022, http://www.etan.org/etanpdf/2006/CAVR/12-Annexe1-East-Timor-1999-GeoffreyRobinson.pdf

5 Soares, S 'Reducing risk in Timor-Leste: a strong foundation for one of the youngest nations', 11 November 2011, *The Conversation*, accessed on 25 September 2022, https://theconversation.com/reducing-risk-in-timor-leste-a-strong-foundation-for-one-of-the-youngest-nations-3839

6 The World Bank Group 2010, *Reducing the Risk of Disaster and Climate Variability in the Pacific Islands: Timor-Leste Country Assessment*, East Asia and the Pacific Region, The World Bank Group, Washington DC. Document No. 53209, accessed 20 September 2021, https://documents1.worldbank.org/curated/en/896231468332385902/pdf/532090WP0P11201MOR1LESTE1ASSESSMENT.pdf

7 Keefer, G, 2000, *Report on the Restoration of Meteorological Network—Timor Loro'Sae. Report II,* United Nations Transitional Administration in East Timor, Dili, East Timor.

8 Barnett, J, Dessai, S,S & Jones, RN 2007, 'Vulnerability to climate variability and change in East Timor', *Ambio,* Vol. 36, No. 5, pp. 372–378.

9 Secretary of State for Environment, República Democrática de Timor-Leste, 2020. *Second National Communication under the United Nations Framework Convention on Climate Change*, accessed on 25 September 2021 *https://unfccc.int/documents/266424*

10 Dolven, B, Margesson, R & Vaughn, B 2012, *Timor-Leste: Political Dynamics, Development, and International Involvement*, accessed on 2 February 2022, http://www.fas.org/sgp/crs/row/R42585.pdf

11 Australian Government 2013, *Current and Future Climate of Timor-Leste. International Climate Change Adaptation Initiative: Pacific Climate Change Science Program,* accessed on 2 February 2022, https://www.pacificclimatechangescience.org/wp-content/uploads/2013/06/5_PACCSAP-Timor-Leste-9pp_WEB.pdf

12 Emergency Capacity Building Project & ACAPS 2012, *Preparedness Timor-Leste: Disaster Needs Analysis*, accessed on 2 February 2022, https://www.alnap.org/help-library/timor-leste-preparedness-disaster-needs-analysis-dna-september-2012

13 The World Bank Group & Asian Development Bank 2021, *Climate Risk Country Profile: Timor-Leste*, accessed on 2 February 2022, https://www.adb.org/publications/climate-risk-country-profile-timor-leste

14 McGuire, B 2012, *Waking the Giant: How Changing Climate Triggers Earthquakes, Tsunamis and Volcanoes,* Oxford University Press, Oxford.
15 Barnett, Dessai & Jones 2007.
16 IFRC Climate Centre (International Federation of the Red Cross and Red Crescent) 2021, *Climate Change Impacts on Health and Livelihoods: Timor Leste Assessment*, accessed on 3 September 2021, https://www.climatecentre.org/wp-content/uploads/RCRC_IFRC-Country-assessments-Timor-Leste-V4.pdf
17 Timor-Leste Secretariat of State for Environment, *Coordinating Minister for Economic Affairs (2021), Timor-Leste's National Adaptation Plan: Addressing climate risks and building climate resilience*, accessed on 10 September 2021, https://www4.unfccc.int/sites/NAPC/Pages/national-adaptation-plans.aspx
18 The World Bank Group and the Asian Development Bank, 2021.
19 UNDP 2018. *National Coastal Vulnerability Assessment and Designing of Integrated Coastal Management and Adaptation Strategic Plan for Timor-Leste: Coastal Vulnerability Assessment Report: National Coastal Vulnerability Assessment and Designing of Integrated Coastal Management and Adaptation Strategic Plan for Timor-Leste* accessed on, 3 September 2021, https://www.tl.undp.org/content/timor_leste/en/home/library/resilience/national-coastal-vulnerability-assessment-and-designing-of-integ.html
20 Dekens, J 2007, *Local Knowledge for Disaster Preparedness: A Literature Review.* International Centre for Integrated Mountain Development, Kathmandu, Nepal
21 IPCC 2022a, *Climate Change 2022: Impacts, Adaptation and Vulnerability*, Contribution of Working Group II to the Sixth Assessment Report of the IPCC, Cambridge University Press, Cambridge, In Press.
22 Hiwasaki, L, Luna, E, Syamsidik & Marçal, JA 2015, 'Local and Indigenous Knowledge on Climate-Related Hazards of Coastal and Small Island Communities in Southeast Asia', *Climatic Change*, Vol. 128, pp. 35–56.
23 Mercer, J, Kelman, I, Rosario, F, de Jesus Lima, A, da Silva, A, Beloff, A-M & McClean, A 2014, 'Nation-building policies in Timor-Leste: disaster risk reduction, including climate change adaptation', *Disaster*, Vol. 36, No. 4, p. 705.
24 Secretary of State for the Environment, República Democrática de Timor-Leste 2021, *Timor-Leste's National Adaptation Plan: Addressing climate risks and building climate resilience*, accessed on 2 February 2022, https://www4.unfccc.int/sites/NAPC/Pages/national-adaptation-plans.aspx
25 Demetrio do Amaral deCarvalho as cited in Rogers, J 2015, 'Ancient traditions protect the environment in East Timor', *Deutsche Welle*, 17 February, accessed 2 April 2022, https://p.dw.com/p/1EcWF
26 Mercer et al 2014.
27 World Bank 2010, p. 18.
28 Sherlock, Kevin, 2000, *Kevin Sherlock Notes*, Timor Archives, accessed on 2 April 2022, https://timorarchives.files.wordpress.com/2011/07/kps-notes.pdf
29 Goglia, F & Afonso, S 2012, 'Multilingualism and language maintenance in the East Timorese diaspora in Portugal', *Ellipsis*, Vol. 10, pp. 97–123, 78.
30 Hajek, J & Goglia, F 2020, 'The East Timorese in Australia: multilingual repertoires, language attitudes, practices and identity in the diaspora', *Journal of Multilingual and Multicultural Development,* Vol. 41, No. 3, pp. 264–81, doi:10.1080/01434632.2019.1617295
31 República Democrática de Timor-Leste 2011, *Timor-Leste Strategic Development Plan 2011–2030, version submitted to National Parliament*, Dili, Timor-Leste.
32 Diaspora Emergency Action & Coordination (DEMAC) 2021, *Diaspora Organizations and Their Humanitarian Response in Timor-Leste: Diaspora Organizations and Their Humanitarian Response in Timor-Leste - Timor-Leste*DEMAC, Copenhagen, Denmark.
33 World Bank 2021, *Learning from Tropical Cyclone Seroja: Building Disaster and Climate Resilience in Timor-Leste*, World Bank, Washington, DC, accessed 15 July 2022, http://hdl.handle.net/10986/36806

34 United Nations Office for the Coordination of Humanitarian Affairs (OCHA), 'Timor-Leste floods: urgent call for assistance', 16 June 2021, https://reliefweb.int/report/timor-leste/timor-leste-floods-urgent-call-assistance
35 Diaspora Emergency Action & Coordination (DEMAC) 2021.
36 *The Northern Territory Times and Gazette,* Palmerston, Port Darwin 1st July 1898, 'Latest from the East', *1 July 1898,* accessed on 12 April 2022, *http://nla.gov.au/nla.news-article4200175*
37 *O Porvir,* 'Tufão em Timor', Hongkong, anno I, no. 30, pp. 1–2, 11 July 1898.
38 CFE-DMHA (Center for Excellence in Disaster Management and Humanitarian Assistance) 2016, 'Timor-Leste: Disaster Management Reference Handbook', *OCHA Reliefweb,* 19 July 2021, https://reliefweb.int/report/timor-leste/timor-leste-disaster-management-reference-handbook-2016
39 Harris, R & Major, JR 2017, 'Waves of destruction in the East Indies: the Wichmann catalogue of earthquakes and tsunami in the Indonesian region from 1538 to 1877' in PR Cummins & I Meilano (eds) *Geohazards in Indonesia: Earth Science for Disaster Risk Reduction,* Geological Society, London, Special Publications, Vol. 441, No. 1, pp. 9–46.
40 F.V., *Boletim da Provincial de Macau e Timor,* Vol. XVI, No. 27, p. 116, 4 July 1870. 'F.V.' was identified by K. Sherlock as João Monteiro Pinto da Fonseca Vaz, an officer onboard the corvette Sá da Bandeira.
41 Center for Excellence in Disaster Management and Humanitarian Assistance (CFE-DMHA) 2016.
42 National Environmental Centers for Environmental Information, National Oceanic and Atmospheric Administration, NCEI Hazard Tsunami Event Information, accessed 2 April 2022, https://www.ngdc.noaa.gov/hazel/view/hazards/tsunami/event-more-info/2245
43 *Indonesian Observer,* 'Earthquake hits East Timor, one dead, 9 missing', 16 May 1995.
44 Nakashima, D, Krupnik, I & Rubis, JT, 2018, *Indigenous Knowledge for Climate Change Assessment and Adaptation,* Cambridge University Press and UNESCO, Cambridge and Paris, p. 71.
45 Harris, R, & Major J 2016 p. 8.
46 World Bank Group 2010, p. 17.
47 IPCC 2022b, *Regional Fact Sheet: Small Islands,* IPCC Sixth Assessment Report: Impacts, Adaptation and Vulnerability, https://www.ipcc.ch/report/ar6/wg2/about/factsheets/

References

Australian Government 2013, *Current and Future Climate of Timor-Leste,* Pacific-Australia Climate Change Science and Adaptation Planning Program, https://www.pacificclimatechangescience.org/publications/country-brochures/

Barnett, J, Dessai, S & Jones, RN 2007, 'Vulnerability to climate variability and change in East Timor', *Ambio,* Vol. 36, No. 5, pp. 372–378. https://www.jstor.org/stable/4315847

Boletim da Provincial de Macau e Timor, 4 July 1870, Vol. XVI, No. 27, p. 116.

CFE-DMHA (Center for Excellence in Disaster Management and Humanitarian Assistance) 2016, 'Timor-Leste: Disaster Management Reference Handbook', *OCHA Reliefweb,* 19 July, https://reliefweb.int/report/timor-leste/timor-leste-disaster-management-reference-handbook-2016

Comissão de Acolhimento, Verdade e Reconciliação (CAVR) de Timor Leste 2005, *Chega! The CAVR Report.* http://www.cavr-timorleste.org/en/chegaReport.htm

Cunneen, J, 2022 '5 ways climate change increases the threat of tsunamis, from collapsing ice shelves to sea level rise', *The Conversation,* 24 January, accessed 2 April 2022, https://theconversation.com/5-ways-climate-change-increases-the-threat-of-tsunamis-from-collapsing-ice-shelves-to-sea-level-rise-175247

Dekens, J 2007, *Local Knowledge for Disaster Preparedness: A Literature Review*, International Centre for Integrated Mountain Development, Kathmandu, Nepal.

DEMAC (Diaspora Emergency Action & Coordination) 2021, *Diaspora organizations and their humanitarian response in Timor-Leste*, DEMAC, Copenhagen, Denmark. https://reliefweb.int/report/timor-leste/diaspora-organizations-and-their-humanitarian-response-timor-leste

Dolven, B, Margesson, R & Vaughn, B 2012, *Timor-Leste: Political Dynamics, Development, and International Involvement*, Congressional Research Service, Library of Congress. https://sgp.fas.org/crs/row/R42585.pdf

Emergency Capacity Building Project & ACAPS 2012, *Preparedness Timor-Leste: Disaster Needs Analysis*, https://www.alnap.org/help-library/timor-leste-preparedness-disaster-needs-analysis-dna-september-2012

Goglia, F & Alfonso, S 2012, 'Multilingualism and language maintenance in the East Timorese diaspora in Portugal', *Ellipsis*, Vol. 10, pp. 73–99. 10.21471/jls.v10i0.89

Hajek, J & Goglia, F 2018, 'The East Timorese in Australia: multilingual repertoires, language attitudes, practices and identity in the diaspora', *Journal of Multilingual and Multicultural Development*, Vol. 41, No. 3, pp. 264–281. 10.1080/01434632.2019.1617295

Harris, R & Major, J 2016, 'Waves of destruction in the East Indies: the Wichmann catalogue of earthquakes and tsunami in the Indonesian region from 1538 to 11877' in PR Cummins & I Meilano (eds.) *Geohazards in Indonesia: Earth Science for Disaster Risk Reduction*, Special Publications, Geological Society, London, UK, pp. 9–46.

Hiwasaki, L, Luna, E, Syamsidik & Marçal, JA 2015, 'Local and indigenous knowledge on climate-related hazards of coastal and small island communities in Southeast Asia', *Climatic Change*, Vol. 128, pp. 35–56. 10.1007/s10584-014-1288-8

IFRC Climate Centre (International Federation of the Red Cross and Red Crescent) 2021, *Climate Change Impacts on Health and Livelihoods: Timor Leste Assessment*, https://www.climatecentre.org/wp-content/uploads/RCRC_IFRC-Country-assessments-Timor-Leste-V4.pdf

Indonesian Observer 1995, 'Earthquake hits East Timor, one dead, 9 missing', 16 May 1995.

IPCC 2022a, *Climate Change 2022: Impacts, Adaptation and Vulnerability*, Contribution of Working Group II to the Sixth Assessment Report of the IPCC, Cambridge University Press, Cambridge, In Press.

IPCC 2022b, *Regional Fact Sheet: Small Islands*, IPCC Sixth Assessment Report: Impacts, Adaptation and Vulnerability, https://www.ipcc.ch/report/ar6/wg2/about/factsheets/

Keefer, G 2000, *Report on the Restoration of Meteorological Network—Timor Loro Sae, Report II*, United Nations Transitional Administration in East Timor, Dili, East Timor.

Liu, C, Linde, A & Sacks, I 2009, 'Slow earthquakes triggered by typhoons', *Nature*, vol. 459, pp. 833–836. 10.1038/nature08042

McGuire, B 2012, Waking the Giant: How Changing Climate Triggers Earthquakes, *Tsunamis and Volcanoes*, Oxford University Press, Oxford.

Mercer, J, Kelman, I, do Rosario, F, de Deus de Jesus Lima, A, da Silva, A, Beloff, AM & McClean, A 2014, 'Nation-building policies in Timor-Leste: disaster risk reduction, including climate change adaptation', *Disaster*, Vol. 36, No. 4, pp. 690–718. 10.1111/disa.12082

Nakashima, D, Krupnik, I & Rubis, JT 2018, *Indigenous Knowledge for Climate Change Assessment and Adaptation*, Cambridge University Press and UNESCO, Cambridge and Paris.

National Oceanic and Atmospheric Administration (NOAA) National Centres for Environmental Information (NCEI), 'NCEI Hazard Tsunami Event Information', https://www.ngdc.noaa.gov/hazel/view/hazards/tsunami/event-more-info/2245

Niksch, L, McHugh, L & Margesson, R 2001, *East Timor Situation Report*, Congressional Research Service, Library of Congress, http://www.globalsecurity.org/military/library/report/crs/RL30975.pdf

The Northern Territory Times and Gazette 1898, 'Latest from the East', 1 July 1898, accessed on 12 April 2022, http://nla.gov.au/nla.news-article4200175

O Porvir 1898, 'Tufão em Timor' [tr. Typhoon in Timor], 11 July.

República Democrática de Timor-Leste 2011, *Timor-Leste Strategic Development Plan 2011-2030, version submitted to National Parliament*, Dili, Timor-Leste. https://www.timorleste.tl/documents/timor-leste-strategic-development-plan-2011-2030/

Robinson, G 2006, *East Timor 1999*: Crimes Against Humanity: A Report Commissioned by the United Nations Office of the High Commissioner for Human Rights (OHCHR), Princeton University Press, http://www.etan.org/etanpdf/2006/CAVR/12-Annexe1-East-Timor-1999-GeoffreyRobinson.pdf

Rogers, J 2015, 'Ancient traditions protect the environment in East Timor', *Deutsche Welle*, 17 February, accessed 2 April 2022, https://p.dw.com/p/1EcWF

Secretary of State for the Environment, República Democrática de Timor-Leste 2020, *Second National Communication under the United Nations Framework Convention on Climate Change*, https://unfccc.int/documents/266424

Secretary of State for the Environment, República Democrática de Timor-Leste 2021, *Timor-Leste's National Adaptation Plan: Addressing climate risks and building climate resilience*, https://www4.unfccc.int/sites/NAPC/Pages/national-adaptation-plans.aspx

Sherlock, K 2000, *Kevin Sherlock Notes*, Timor Archives, https://timorarchives.files.wordpress.com/2011/07/kps-notes.pdf

Soares, S 2011, 'Reducing risk in Timor-Leste: a strong foundation for one of the youngest nations', *The Conversation*, 11 November, https://theconversation.com/reducing-risk-in-timor-leste-a-strong-foundation-for-one-of-the-youngest-nations-3839

World Bank 2010, *Reducing the Risk of Disasters and Climate Vulnerability in the Pacific Islands: Timor-Leste Country Assessment*, World Bank, Washington, D.C. https://documents1.worldbank.org/curated/en/896231468332385902/pdf/532090WP0P11201MOR1LESTE1ASSESSMENT.pdf

World Bank & ADB (Asian Development Bank) 2021, *Climate Risk Country Profile: Timor-Leste*, World Bank, Washington, D.C. https://www.adb.org/publications/climate-risk-country-profile-timor-leste

World Bank 2021, *Learning from Tropical Cyclone Seroja: Building Disaster and Climate Resilience in Timor-Leste*, World Bank, Washington, D.C. http://hdl.handle.net/10986/36806

United Nations Office for the Coordination of Humanitarian Affairs (OCHA), 'Timor-Leste floods: urgent call for assistance', 16 June 2021, https://reliefweb.int/report/timor-leste/timor-leste-floods-urgent-call-assistance

UNDP (United Nations Development Plan) 2018, National Coastal Vulnerability Assessment and Designing of Integrated Coastal Management and Adaptation Strategic Plan for Timor-Leste: Coastal Vulnerability Assessment Report, https://www.tl.undp.org/content/timor_leste/en/home/library/resilience/national-coastal-vulnerability-assessment-and-designing-of-integ.html

6

CHANGED RESPONSES TO THE CHANGING THREAT OF CLIMATE-INDUCED FIRE AND DROUGHT

Marcelle Scott and Robyn Sloggett

Two climatic extremes that result from the current climate crisis are increased precipitation resulting in floods and decreased precipitation resulting in droughts. Focusing on the impacts of wildfire as outcomes of this cycle, the chapter examines the impacts of 'one in a hundred year events', which are now occurring on almost an annual basis. It explores how traditional knowledge, as a form of intangible heritage, is being used to help address this risk. While droughts produce the tinder-dry country that increases the risk of fire, it is wet springs and summers that increase the fuel load which is necessary to fuel wildfires. Increased and increasingly unpredictable weather events, and the flood and fires that result from such events, reduce community resilience and diffuse the limited resources that are available for recovery. They often result in large-scale catastrophic loss, including the sites, buildings and objects that are important for a community both for day-to-day functioning and for providing social meaning and continuity. Such catastrophic events are a particular issue for heritage experts, as heritage can often be seen as the discretionary, 'nice to do but not critical' part of planning, mitigation and response processes.[1]

Three of the recurring themes found in this book are leitmotifs in this chapter. The first is that heritage is a valuable resource, providing knowledge about how communities have managed, learnt from and adapted to changes in their environment. The second is that heritage provides valuable data that can help to predict the local impacts of climate change-related disasters. The third is the role of Indigenous knowledge as a critical part of the heritage that is being brought to bear on climate change-induced emergencies.

In 2020, eastern Australia experienced what was described as the worst drought for more than a century. At the same time, the western United States was also experiencing catastrophic droughts that were so severe that the Smithsonian Magazine led with the headline: 'The Western USA is Experiencing the Worst

DOI: 10.4324/9781003163312-7

Drought in More Than 1,200 years'.[2] Between the east coast of Australia and the west coast of the United States lies the Pacific Ocean, the largest and deepest part of the globe, covering 165.2 million square kilometres (60 million square miles) with an average depth of 4,000 m (13,000 feet). It is both a heat sink and a producer of moisture. Episodic environmental shifts across the Pacific cause changes to weather patterns and local environments affecting the large land masses that border the Pacific and the small islands that exist in the Pacific. Changes in the distribution of heat and wind across the Pacific are caused by the fluctuating cycles of El Niño and La Niña, creating cyclic events that alter rainfall patterns over a number of years.[3] Researchers in China and the United States examined 33 El Niño events from 1901 to 2017 to ascertain changes in the El Niño and La Niña cycle. They concluded that 'if the observed background changes continue under future anthropogenic forcing, more frequent extreme El Niño events will induce profound socioeconomic consequences'.[4] If this is the case, then environments across the Pacific, the Americas, Australia, Southeast Asia and the Antarctic will be affected by increasingly hostile weather extremes. How we learn to manage these extremes, and what kinds of adaptation might be needed, are highly significant questions.

Australian bushfires in the summer of 2009

In the early summer of 2009 the State of Victoria, in southeastern Australia, was on high alert. Melbournians were experiencing 'the driest start to a year on record'.[5] The previous ten years had been the driest since 1900.[6] The high temperatures and low rainfall throughout the decade-long 'Millennium Drought' were profoundly affecting Australia's environment, economy and national psyche. In rural areas, farmers and communities were under severe stress, mentally, physically, emotionally and economically. Plants, animals and wildlife habitats were impacted. As a result, the attitude of Australians to the idea of a climate crisis was changing.[7] Australia's national science agency, the Commonwealth Scientific and Industrial Research Organisation (CSIRO) forecast further declines in water availability in Australia's major river systems. Many Australians had begun to believe 'their country was a "canary in the coal mine" when it came to climate change and that drought conditions were an early indication of changes that other regions of the world are likely to experience in the future'.[8]

In Australia, mid-summer occurs around mid-January. In 2009, record high temperatures throughout January resulted in a predicted fire danger level reaching unprecedented levels. There were no indicators on the scale of risk that could measure this level of fire danger.[9] On Friday 6 February 2009 the Victorian Premier, John Brumby, warned that the weather forecast for the weekend was foreboding, and that the combination of extraordinarily high temperatures and extraordinarily high winds presented the 'greatest fire danger in the State's history'.[10] He emphasised that conditions were 'tinder-dry' and reiterated the advice of the Chief Fire Officer, that 'it will be as bad as you can get'.[11] Tragically, the next day, soon to become known as Black Saturday, proved to be Australia's deadliest bushfire event. The Black Saturday bushfires, burnt for over a month. They were finally brought under control in mid-March, but not before 173 people had died and another 44 people had been severely injured. The cost to infrastructure, livestock and the natural environment was horrendous. Over 5000 buildings and houses were damaged or destroyed, two entire townships were obliterated, with damage to infrastructure and the natural environment occurring across 450,000 hectares. The fires had devasting consequences for both the social and physical fabric of thousands of people's lives.[12] This was a day that would re-write Australian climate and fire history and shock the public into the realisation that the bushfire risk was part of 'climate-changed meteorological conditions'.[13]

On Monday 9 February Australia's then Deputy Prime Minister Julia Gillard, choking back tears, opened Parliament with a condolence message on behalf of all Australians, 'to those families suffering most, to the communities lost and to a State that will never be the same'.[14] She spoke of the depths of the tragedy that was 'beyond belief, beyond precedent, beyond words' and of the social impact on, 'the beautiful towns and hamlets of Kinglake, Marysville and Narbethong [that] are no more' and the hundreds of 'homes and their irreplaceable contents – like the photographs of children and memories of family life – [that] have been destroyed'.[15]

In 2012, seeking a word to describe the particularly deep emotions people feel when they are faced with the human-induced desolation of a 'once predictable loved home environment', Glenn Albrecht coined the term 'solastalgia'.[16] Albrecht, an environmental philosopher at the University of Newcastle in the Upper Hunter Region of New South Wales, was well known in the community as an advocate for the environment. The area has a long history of coal mining with associated environmental and human health impacts. Local residents would contact Albrecht about environmental concerns and share their 'distress about the threats to their identity and well-being'.[17] Albrecht was interested in the relationships between humans and nature, and was familiar with David Rapport's term 'eco distress syndrome.'[18] Elyne Mitchell's 1946 book *Soil and Civilisation* introduced him to the term 'psychic stability', a term he credits with triggering his interest in the connection between human health and the quality of the environment. Recognising that the symptoms of distress that locals were describing as due to a 'land health-human health' disconnect, Albrecht set out to find a suitable term to describe their feelings. Derived from the concepts of solace and desolation, Albrecht defines solastalgia as

> the pain experienced when there is recognition that the place where one resides and that one loves is under immediate assault (physical desolation). It is manifest in an attack on one's sense of place, in the erosion of the sense of belonging (identity) to a particular place and a feeling of distress (psychological desolation) about its transformation.[19]

These kinds of feelings of grief were reported by many of the 1,000 participants interviewed anonymously for the study 'Beyond Bushfires: Community,

Resilience, and Recovery' which examined the impacts of the 2009 Victorian bushfires 'on the mental health and wellbeing of community members, with a focus on how individual outcomes were influenced by social connections and community-level recovery'.[20] Most participants identified a deep personal connection to the natural environment, and had experienced 'profound feelings of loss and grief at damage to the natural landscape'.[21] Some found solace in observing the regeneration of the landscape while for others the sight of dead animals and the damaged landscape brought back overwhelming feelings of grief.[22] As onc participant put it:

> It was hard, driving through, it was pretty hard ... I could not drive up or down without just, on my own, just crying, just bizarre. Just driving through the place ... and this was for months ... totally 'nuked, ... but it's still what you remember.[23]

Collections, museums and bushfires

Museums tell stories through the things they collect. By providing a sense of collective heritage, they help to keep memories alive, and assist individuals to understand who they are, and their place in the wider world. In many small communities, they are the collective memory of place. There is any number of professional guides with recommendations on how best to protect collections in times of disaster. Aware of the impending danger in the 2009 bushfire season, the Marysville District Historical Society, located in the rugged, mountain ash rainforest in northeast Victoria, had implemented most of these.

Like many small district historical societies, Marysville District Historical Society is managed by local volunteers. Facing a catastrophic summer, they moved the collection to three different locations to reduce the risk, a sensible plan that is clearly identified as a strategy in professional guides. The 2009 fires destroyed most of Marysville including the places designated as safe havens for the collection. One hundred and forty-five years of heritage were lost.[24] The Society quickly issued a public call for donations of historical records relating to the Marysville region, and the Marysville District Historical was quickly in receipt of thousands of local and international donations. Within 16 months of the fires, the dedicated volunteer curators Mary and Reg Kenealy had secured the funds to purchase a new building, and began moving the new collection in, smiling and confident 'that the stories of Marysville and district can still be told'.[25]

When so much is lost, even damaged or apparently insignificant items can be precious. This was the case in the aftermath of the 1983 fires in Macedon, Victoria, documented in Peter Read's book *Returning to Nothing: The Meaning of Lost Places*. In his interview with a woman who had lost everything, she recalled her feelings when the tractor crew arrived to demolish the brick chimney, the only part of her house left standing. 'I was so angry and I went out there like a mad woman screaming, "leave it alone". That was all we had, just a chimney'. With everything else lost, this chimney had acquired new meanings. It was 'the site of collective family memories, a symbol of the past, a reminding of a terrifying trauma and a protector of their possessions'.[26]

Shortly after the Victorian fires were extinguished Museum Victoria established its Victorian Bushfire Collection which sought 'to preserve and document the impact of the Black Saturday fires, the community response, the aftermath and the process of recovery and renewal'.[27] The Museum hoped that the project would actively engage individuals who could participate, commemorate and share stories from their bushfire experiences. The Museum also hoped that the Collection would 'foster greater public understanding of bushfires, how they have shaped the history and landscape of Victoria and will continue to do so'.[28]

Conscious of the emotional trauma and exhaustion that people affected by the fires were experiencing, members of the Australian Institute for the Conservation of Cultural Material (AICCM) decided the most useful first step was to circulate information about how to salvage their most meaningful possessions safely and effectively. As Alex Ellem, then President of the Victorian Branch of the AICCM noted:

> We knew that as the fires had caused such destruction to lives and communities, the last thing people would really be thinking about was their surviving possessions. At the same time, we knew that if people were able to retrieve possessions from their homes, salvaged items would take on a new significance in their lives, being in many cases the sole material connections to their past.[29]

The information was shared on the AICCM website, on discussion lists, with regional councils in affected areas, and via a joint press release with the Heritage Council of Australia with the 'immediate aim … to inform people about their personal safety and how to handle and care for the salvaged items' once they were ready.[30] Some of the councils expressed their interest in hosting visits from conservators. Others, while interested, felt it was too soon, with many communities 'already suffering meeting fatigue, having been addressed by a seemingly never-ending array of service providers and emergency service personnel, while dealing with infinite grief and trauma'.[31] Balancing support for affected communities while recognising the pressures they are under means that recovery from a disaster requires a long-term, coordinated response.

Responding to the 2019 Australian bushfires

A little over ten years after the 2009 Victorian bushfires, Australia would once again experience a catastrophic bushfire season. 2019 was Australia's hottest and driest year on record, and the national Forest Fire Danger Index was the highest since records began.[32] Professor Ross Garnaut, who had conducted a review of the impacts of climate change on the Australian economy in 2008, predicted that

Australia's bushfire seasons would start earlier and end later. Tragically, he was right, and bushfires burned out of control from July 2019 until they were finally extinguished in March 2020.[33] Now colloquially referred to as 'Black Summer', high-severity fires raged across millions of hectares in every state of Australia, with hundreds of communities threatened and displaced as a result.[34] The impact was massive. Eighty percent of the Australian population was affected by smoke and associated poor air quality with between 650 million and 1.2 billion tonnes of carbon dioxide released into the atmosphere.[35] By the time the fires were finally brought under control 34 people had died. Four thousand people were admitted to hospital due to complications directly attributable to the smoke, and another 445 people would die from its effects. Around 9,000 homes and outbuildings and over 100 of the State's most significant historical heritage sites were razed or badly damaged. An estimated 3 billion animals, including many of Australia's endemic and critically endangered native species, had been killed, injured, or had their habitat destroyed.[36] Dermot O'Gorman, then CEO of the Australian branch of the World Wide Fund for Nature described Black Summer 'as one of the worst wildlife disasters in modern history'.[37]

Graphic images of exhausted firefighters, burnt-out landscapes, dead and injured animals and the blackened remains of vehicles, houses and their contents dominated television and newspaper bulletins across the country.[38] On 20 February 2020 Prime Minister, Scott Morrison, announced that there would be a Royal Commission into National Natural Disaster Arrangements.[39] With at least one public inquiry into bushfires every two years since 1940 and with many recommendations still not implemented, fire experts questioned the need for yet another.[40] Peter Marshall, National Secretary of the United Firefighters Union of Australia and experienced firefighter, felt that another 'rehashing of old recommendations only to gather dust on parliamentary shelves' was 'the last thing we need, at the end of a difficult bushfire season'.[41] He recommended instead that the lessons already learnt from previous enquiries should be re-examined.[42] Nonetheless, the Royal Commission into National Natural Disaster Arrangements went ahead, chaired by Air Chief Marshall Mark Binskin AC (Retd).

The Commissioners published their wide-ranging multi-volume report on 28 October 2020. *Recommendation 15.3: Prioritising Mental Health During and After Natural Disasters* responds to numerous submissions and witness statements that refer to the immediate and the long-term pervasive effects of traumatic stress, anxiety, depression and Post Traumatic Stress Disorder (PTSD) often experienced by first responders and in many cases their partners and families.[43] Chapter 16 of the report, *Wildlife and Heritage*, recognises the interconnecting social and cultural values of environment and heritage, and the need for an integrated approach to emergency management. The Commissioners acknowledged that this 'requires accessible data, including on the location of environmental, heritage and cultural sites, the distribution of species and ecological communities and priorities to guide response efforts'.[44]

Royal Commissions have broad investigative powers, with responsibility to report to governments on what needs to change. Despite criticism of Royal

Commissions as time consuming, expensive and focussed on blame, they provide a rare opportunity for individuals and groups to air their views and experiences in the knowledge that their words and opinions will be heard, treated respectfully, anonymously if they wish, and committed to the public record, and 'give voice to victims'.[45] During 35 days of hearings, the Binskin Royal Commission considered more than 1,750 written submissions, reviewed nearly 80,000 pages of documents and heard from more than 270 witnesses. One respondent described the sense of community spirit, strength and resilience as the long-term impact on her personally and on the community in which she lives. Another focussed her concerns on 'the significant mental health impacts that will have an everlasting flow-on effect upon thousands, including intergenerational trauma to many children, well into the future'.[46] Others were unable to respond, being 'too psychologically injured or impaired to be able to relive the difficulties they have faced or to articulate how they have been suffering'.[47] A frighteningly large number of people remained living in tents as winter approached. This is the harsh reality that should be borne in mind, when engaged in any stage of disaster planning, mitigation, response and post-disaster recovery.

Denis Rose, Gunditjmara Elder and Traditional Owner of the lands on and surrounding the World Heritage-listed Budj Bim cultural landscape in southwest Victoria, advocated for the incorporation of Traditional Knowledge of land and fire management into mainstream firefighting practices.[48] This point was also identified elsewhere in the report. In Chapter 18, *Indigenous Land and Fire Management,* the Commissioners describe the multiple benefits to Indigenous, non-Indigenous, and other local community knowledge holders from the use of Aboriginal fire management systems. *Recommendation 18.1: Indigenous Land and Fire Management and Natural Disaster Resilience* states that 'Australian, state, territory, and local governments should engage further with Traditional Owners to explore the relationship between Indigenous land and fire management and natural disaster resilience'.[49]

The Commissioners found that:

> Indigenous land management is an example of how local knowledge has successfully informed land management, and it has done so for tens of thousands of years. Indigenous land management draws on a deep knowledge of Australia's landscapes. It is based on cultural understandings of Country, is tailored to specific places, and engages local people in development and implementation. Partly for these reasons, Indigenous land management differs widely across Australia.[50]

Indigenous knowledge and Indigenous site management

The Budj Bim Cultural Landscape received UNESCO World Heritage Listing in 2019. With evidence of 6,600-year-old stone dwellings and lava flows adapted to form one of the oldest aquaculture sites in the world, it was the first Indigenous Australian landscape to be listed for purely cultural values. In this millennia-old landscape, fire is no stranger. In January 2020, fire burnt through more than 7,000 hectares in the Budj Bim National Park including parts of the heritage site. Previous collaborations between the local fire authorities and the Gunditj Mirring Traditional Owners identified that low-impact fire control techniques would be useful if there was an outbreak, both to protect the site and to be able to traverse the rocky landscape.

As program manager of the Gunditj Mirring Traditional Owners Aboriginal Corporation, Denis Rose described how the Weeyn Yakeen (fire dreaming) program was critical to the protection of the site during the fires. Weeyn Yakeen is a collaborative fire management strategy in partnership with the CFA (Country Fire Authority, now Fire Rescue Victoria), the Department of Environment, Land, Water and Planning (DELWP) and the Gunditjmara Aboriginal Corporation that develops three-year cool burning plans. For Rose, the key message from his experience with the fires

> has been the great relationships we have as traditional owners … we've worked on these for a long time. … So we have these robust partnerships and we have a lot of trust and confidence in each other that we can … sometimes make the hard decisions. As I said, the fire controllers, they had to make a hard decision about what fire suppression method they could use. They did listen to our concerns and our advice, and it worked.[51]

The story of the fires at Budj Bim is a positive one of Indigenous history, heritage, leadership, and multi-agency collaboration. A study by Williamson, Markham, and Weir tells a different story. Their research draws on published material to 'demonstrate that the Aboriginal experience of the 2019–2020 bushfires have been different from those of non-Indigenous Australians'.[52] A selection of these accounts illuminates this difference and reveals another social, cultural, and health impact of the fires, that of 'ecological grief'. This is best described as grief arising from climate change-related loss of country, species and ecosystems. For Indigenous people across the colonised world loss of and changes to country, it is a trauma that is, tragically, re-occurring.[53]

Amy Thunig, a Gomeroi/Gamilaroi/Kamilaroi scholar and writer identifies a connection between the impacts of the bushfires and those of colonisation in her emphatic statement:

> Seeing people start to ask, 'what is going on in Australia?!' Ongoing colonisation, that's what's going on. This land was nurtured in reciprocal relationships with First Nations peoples for some 120,000 + years. THIS is the result of 200 odd years of extractive invasion.[54]

Lorena Allam a Gamilaraay and Yawalaraay journalist was reporting on the fires to *The Guardian* newspaper and gave a personal message, critical of those who deny climate change.

> Like you, I've watched in anguish and horror as fire lays waste to precious Yuin land, taking everything with it – lives, homes, animals, trees – but for First Nations people it is also burning up our memories, our sacred places, all the things which make us who we are. It's a particular grief, to lose forever what connects you to a place in the landscape. Our ancestors felt it, our elders felt it, and now we are feeling it all over again as we watch how the mistreatment and neglect of our land and waters for generations, and the pig-headed foolishness of coal-obsessed climate change denialists turn everything and everyone to ash.[55]

For Vanessa Cavanagh, a Bundjalung and Wonnarua woman, the loss of a Grandmother tree that she had grown up with on Darkinjung Country was personal:

> The sight of this old tree with her crown removed brought warm, stinging tears to my eyes. It was a deep hurt of losing someone far older and wiser than me. Losing someone who was respected and adored. Someone with knowledge I cannot fathom or comprehend. When I told my mum that evening she reacted similarly, a personal and family loss. To others she might just be a big tree.[56]

Victor Steffensen, a Tagalaka man, Indigenous fire practitioner and author, expressed his frustration at the continued exclusion of Aboriginal people in bushfire planning and management. Steffenson was a panellist on the live Australian television programme *Q&A Bushfires Special* broadcast on 2 February 2020 in the wake of the 2019–2020 fires. When asked 'What would you say to the authorities that we could do ahead of next summer?' He replied:

> I would say jump in the passenger seat and let us drive for a change. And when you're in the passenger seat of a car, that means we don't leave you behind. And it's too often that we're never even in the passenger seat and we get left behind all the time. And just once in this nation's history can you just listen to Aboriginal people, our knowledge system? You know, there's an intelligence there and we have all this information for looking after the environment, and we're not being tapped into, and it's so frustrating.[57]

Noel Webster, a Yuin man and cultural fire practitioner, experienced the devastation of his country from the 2019–2020 bushfires. He acknowledges the frustration and grief in his community, and uses his experience to inspire future generations.

> It's hard, you do see a lot of heartbreak in the country, you feel that emotion … what is pushing me along is that I've been working with a lot of young people. I've been seeing the effort they put in, their commitment, their

> understanding and it makes my job a lot easier. We're investing a lot in these young fellas who are stepping up and making everyone proud.[58]

Webster knows from experience how successful cool burning practices can be and urges a long-term collaborative commitment across the country that will merge knowledge systems to rebuild and repair the landscape. Achieving this will require more than short-term answers. As Webster says, 'We need investment into Indigenous knowledge, there's no good acknowledging it then walking away. We need actions and outcomes … We have fire practitioners across the country, have a listen and let us have a go'.[59]

The Australian Red Cross 2020 report on the bushfires, *The Power of Humanity*, uses the stories of those who worked with, or were assisted by Red Cross volunteers and staff, before, during and after the fires. Jenny Brown, a Wandi Wandian woman of the Yuin people of the south coast of NSW, and Head of Aboriginal and Torres Strait Islander Strategy for the Australian Red Cross describes her personal experience of the fires and their long-lasting impacts:

> I witnessed firsthand the damage caused to lives and property. It was devastating and incomprehensible. I listened to my people talk of how ferocious and scary the fires were. But it's not just what people said that was frightening. It's what we didn't hear; not a sound from nature. No birds, no wildlife.
>
> I couldn't begin to understand the impact on our sacred sites, songlines, our totems, artefacts, bush foods and medicines. Our totems protect us, and we must protect them; when they or sacred sites are destroyed, it tears our soul and severs our spirit.
>
> I wondered about the harm to my totem the black cockatoo. On that day I travelled to the area, a black cockatoo landed on a tree nearby and looked directly at us. To me, it was a sign of hope, spiritual connection and strength.
>
> Our traditional relationship and interaction to the land are integral to our health, wellbeing and culture, and it will be a long road to recovery and healing for my people.[60]

Walbunja Yuin woman Sherrie, an artist and an environmental ranger with Mogo Land Council, on the south coast of New South Wales, had her home damaged in the fires. While she talks of her fear during the fires, she also describes the strength she gained from talking to Elders and working together. She asserted that

> not many things have been able to scare me in my lifetime. But this fire, she came with one helluva night and she showed us all how vulnerable we really are, … and then as a community, we all just started helping each other out … that's the way communities are supposed to be.[61]

Having done so much to protect the environment both privately and in her role as a ranger, witnessing the damage to the environment from the fires had a 'huge toll' on her mental health. 'It took a long time for me to come to terms with the fact that I do my best for the environment but I couldn't stop a catastrophe like this'.[62] After the fires, Sherrie returned to work as a ranger, which she says has allowed her to show 'the kids, and others in the community who look up to me, that the fire knocked me down but I'm up and walking now'.[63]

In the 2022/23 Australian Federal Budget, the Indigenous Rangers 'Caring for Country' program received a $636 million commitment to create 1000 full-time equivalent workers. The Minister for Indigenous Affairs claimed:

> It will address the unmet demand for cultural burning and other Indigenous land-management expertise in areas outside of the current ranger footprint and also help Australia meet a range of international environmental commitments.[64]

Communicating the climate change crisis

This chapter is about the environmental and social impacts of climate change, as manifested in increasing severe bushfire activity. As the stories in the chapter demonstrate working in bushfire preparation and recovery is complicated and intense. While the stories here focus on the experiences of two recent bushfire seasons in southeast Australia, bushfires (or wildfires or forest fires, they are much the same) are becoming an increasing concern in Europe, Asia, the Middle East, the Americas and Africa.[65] In all cases, communicating the risks to cultural heritage from climate-change-induced fire, and identifying the value of cultural heritage in the fight against climate change, are valuable, complex tasks that require co-ordination, commitment and funding. When the fires are active, heritage professionals may not be able to make contact with affected communities, and this can continue for months after the fires. When it is possible to make contact, professionals are often intervening in the lives of deeply traumatised individuals. Knowledge about local heritage is diverse and is spread across professional, community and cultural organisations and held by individuals. Knowledge about mitigation and response is similarly diversely distributed and may be difficult to access during and after the crisis. For these reasons, it is important that heritage professionals are proactive in communities, and support local systems prior to any crisis so that this knowledge can be activated as required for disaster management, mitigation and response.

A review of conservation literature and conservation heritage literature indicates a reliance on formal language and structured plans that sit at odds with the stressed and emotional accounts of bushfire experience in the various Royal Commission reports cited above. Some advise getting a Heritage Officer to undertake an assessment rather than the owner, but in a traumatised community the site of strangers making notes on clipboards can be confronting.[66]

Recognising the importance of sharing personal stories about the impacts of disasters and the need for clear, accurate, and plain language communication about the links between climate change and severe weather events the Climate Council of Australia has identified a set of 'golden rules' for climate communications. These are:

> Talk about problems that are local, immediate and relevant to people's lives.
>
> Highlight how widespread concern for climate change is … demonstrate that action is already underway across communities and constituencies of all types.
>
> Outline a positive vision of the future … how renewable energy and storage can transform electricity systems … provide … cheap, clean and abundant power and increase resilience during extreme weather.
>
> Remember the importance of the messenger as well as the message and equip those who have high credibility with the tools they need to speak out.[67]

The page dedicated to Bushfires focuses on increased risk, the difficulty of planning, and provides a clear and simple outline of the ways in which climate change makes bushfires worse.[68]

The Climate Council's 'Communication Guide' also suggests focussing on 'People impacted by the bushfires, such as individuals sheltering in a community refuge, escaping the danger, or inspecting their impacted property'.[69] These guides are available in Mandarin, Vietnamese, Cantonese, Arabic, Korean, Greek, Thai, Italian, Khmer and Spanish. In addition to the Communication Guides, the Climate Council's website also provides multilingual social media graphics (Infographics) to assist people wanting to provide public advice or public statements about how climate change affects bushfires and catastrophic floods (Figs. 6.1 and 6.2). Such resources enable clear and consistent messaging in complex, emotive debates about the role of climate change in extreme bush fires.

The range of heritage experts identified in this chapter includes Indigenous Traditional Knowledge Holders, Indigenous Traditional Owners, volunteers, community members and heritage professionals, many of whose roles overlap across these communities of experts. Their stories and their roles are diverse, yet they overlap in very important ways as all of these groups have knowledge and experience in bushfire mitigation, response and recovery. Heritage professionals sit in one small vector of this tessellated body of knowledge. Nevertheless, in addition to providing advice on how to document and to safely store and recover cultural heritage damaged by fire, heritage professionals can play a significant role in leading the climate change debate. Heritage experts can point to the need for communities to take responsibility for their cultural heritage in the face of unprecedented risk.

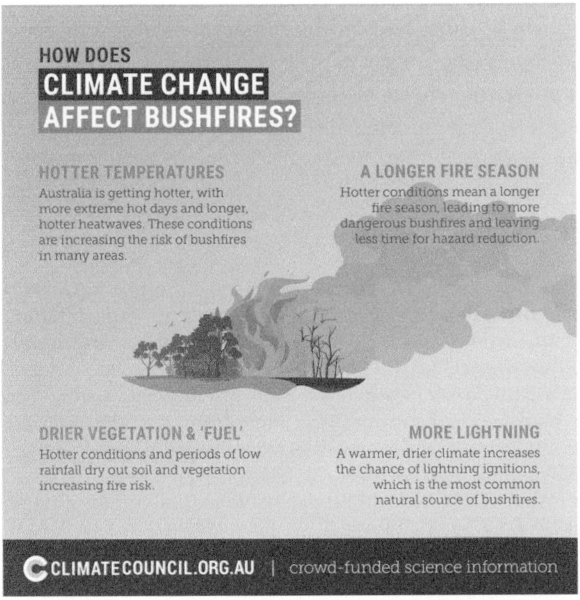

FIGURE 6.1 Climate Council social media graphic 'How does climate change affect bushfires?'

Licensed under Creative Commons - Climate Council of Australia Ltd

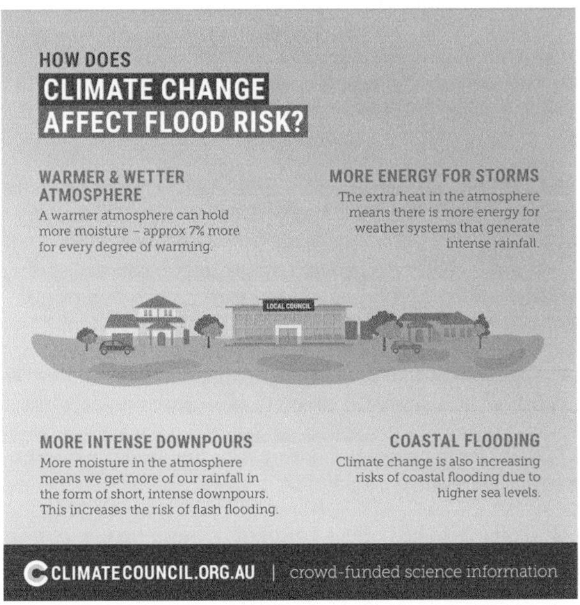

FIGURE 6.2 Climate Council social media graphic 'How does climate change affect flood risk?'

Licensed under Creative Commons - Climate Council of Australia Ltd

To conclude, as the experience of those affected by bushfires shows, clear communication about the links between the current climate crisis and risks to heritage is critical in any attempts to mitigate the effects of climate change-induced fires. The links between cultural heritage experts, sustainability and adaptation considered in more detail in the following chapters.

Notes

1 Brown, D 2014, 'Preservation in the 21st century: preservation is a political movement', *National Trust for Historic Preservation*, accessed on 8 April 2021, https://forum.savingplaces.org/blogs/forum-online/2014/07/16/preservation-in-the-21st-century-preservation-is-a-political-movement
2 Osborne, M 2022, 'The Western USA is experiencing the worst drought in more than 1,200 years', *Smithsonian Magazine*, 17 February 2022, https://www.smithsonianmag.com/smart-news/the-western-us-is-experiencing-the-worst-megadrought-in-more-than-1200-years-180979590/
3 Carlowicz, M & Schollaert Uz, S 2017, 'El Niño pacific wind and current changes bring warm, wild weather', *NASA Earth Observatory*, 14 February, accessed on 2 April 2022, https://earthobservatory.nasa.gov/features/ElNino
4 Wang, B, Luo, X, Yang, Y, Sun, W, Cane, MA, Cai, W, Yeh, S & Liu, J 2019, 'Historical change of El Niño properties sheds light on future changes of extreme El Niño', *Proceedings of the National Academy of Sciences of the United States of America*, Vol. 116, No. 45, pp. 22 512–22 517.
5 Bureau of Meteorology (BOM) 2019a, 'Remembering Black Saturday: the extraordinary weather behind Victoria's 2009 bushfires', *BOM Blog*, blog post, accessed on 26 February 2021, https://media.bom.gov.au/social/blog/2025/remembering- black-saturday-the-extraordinary-weather-behind-victorias-2009-bushfires/
6 Bureau of Meteorology (BOM) 2009, 'Special climate statement 14', Australian Government, p. 3, http://www.bom.gov.au/climate/current/statements/scs14.pdf
7 Heberger, M 2012, 'Australia's millennium drought: impacts and responses' in PH Gleick (ed) *The World's Water, Volume* 7, Island Press, Washington, DC, pp. 97, 102.
8 Heberger 2012, p. 102.
9 BOM 2019a.
10 ABC News 2009, '"Worst ever" fire risk threatens Vic', 6 February, accessed on 18 October 2021, https://www.abc.net.au/news/2009-02-06/worst-ever-fire-risk-threatens-vic/285892
11 Moncrief, M 2009, '"Worst day in history": Brumby warns of fire danger', *Sydney Morning Herald*, 6 February, accessed on 24 February 2021, https://www.smh.com.au/national/worst-day-in-history-brumby-warns-of-fire-danger-20090206-7zf1.html
12 Gibbs, L, Molyneaux, R, Harms, L, Gallagher, C, Block, K, Richardson, J, Brandenburg, V, O'Donnell M, Kellett, C, Quinn, P, Kosta, L, Brady, K, Ireton, G, MacDougall, C & Bryant, R 2020, *10 Years Beyond Bushfires Report*, University of Melbourne, Melbourne, Australia.
13 O'Neill, SJ & Handmer, J 2012, 'Responding to bushfire risk: the need for transformative adaptation', *Environmental Research Letters*, Vol. 7, p. 2.
14 Gillard, J 2009, *Condolence Speech - Victorian Fires*, Ministers' Media Centre Dept. of Education, Skills and Employment, accessed on 14 May 2021, https://ministers.dese.gov.au/gillard/condolence-speech-victorian-fires
15 Gillard 2009.
16 Albrecht, G 2012, 'The age of solastalgia', *The Conversation*, 7 August, accessed on 17 May 2021, https://theconversation.com/the-age-of-solastalgia-8337

17 Albrecht, G 2005, 'Solastalgia: a new concept in human health and identity', *PAN (Philosophy, Activism, Nature)*, Vol. 3, p. 41.
18 Rapport, D & Whitford, WG 1999, 'How Ecosystems Respond to Stress', *Bioscience*, Vol. 49, No. 3, pp. 193–203.
19 Albrecht 2005, p. 45.
20 Gibbs et al 2020, p. 7.
21 Gibbs et al 2020, p. 19.
22 Block, K, Molyneaux, R, Gibbs, L, Alkemade, N, Baker, E, MacDougall, C, Ireton, G & Forbes, D 2019, 'The role of the natural environment in disaster recovery: "We live here because we love the bush"', *Health & Place*, Vol. 57, p. 61.
23 Block et al 2019, pp. 64–65.
24 Webb, C 2009, 'Marysville's past is pouring back in', *Sydney Morning Herald*, August 7, accessed on 29 October 2021, https://www.smh.com.au/national/marysvilles-past-is-pouring-back-in-20090806-ebk2.html
25 Webb, C 2010, 'Marysville's history rises from the ashes to find new home', *The Age*, 7 June, accessed on 29 October 2021, https://www.theage.com.au/national/victoria/marysvilles-history-rises-from-ashes-to-find-new-home-20100606-xna7.html
26 Read, P 1996, *Returning to nothing: The Meaning of Lost Places*, Cambridge University Press, Cambridge, p. 108.
27 Dale-Hallett, L 2009, 'Victorian bushfires collection', *Museums Victoria*, accessed on 29 October 2021, http://collections.museumsvictoria.com.au/articles/3032
28 Dale-Hallett 2009.
29 Ellem, A 2011, 'In support of public and private treasures: the AICCM Victorian Division's disaster response to the Victorian fires of 2009', *AICCM Bulletin*, Vol. 32, No. 1, p. 191.
30 Ellem 2011, p. 191.
31 Ellem 2011, p. 192.
32 Binskin, M 2020, *Royal Commission into National Natural Disaster Arrangements: Interim Observations*, Commonwealth of Australia, accessed on 15 November 2020, p. 5; BOM 2019b, *Tracking Australia's climate through 2019*, accessed on 1 July 2021, p. 19, http://www.bom.gov.au/climate/updates/articles/a036.shtml
33 Garnaut, R 2008, *The Garnaut Climate Change Review Final Report*, Cambridge University Press, Port Melbourne, Australia.
34 Binskin 2020, p. 19.
35 Hughes, L, Steffen, W, Mullins, G, Dean, A, Weisbrot, E & Rice, M 2020, *Summer of Crisis*, Climate Council of Australia, p. iii, https://www.climatecouncil.org.au/resources/summer-of-crisis/
36 Binskin 2020, p. 5.; Hughes et al 2020, p. 11.
37 O'Gorman, D 2020, 'Forward' in World Wildlife Fund for Nature Australia 2020, *Australia's 2019–20202 Bushfires: the wildlife toll*, accessed on 6 May 2021, p. 1, https://www.wwf.org.au/news/news/2020/3-billion-animals-impacted-by-australia-bushfire-crisis#gs.e6ic8d
38 Burgess, T, Burgmann, JR, Hall, S, Holmes, D & Turner, E 2020, *Black Summer: Australian Newspaper Reporting on the Nation's Worst Bushfire Season*, Monash Climate Change Communication Research Hub, Monash University, Melbourne, p. 5.
39 Morrison, S 2020, 'National Royal Commission into Black Summer bushfires established' [press release], Prime Minister of Australia, accessed on 17 July 2021, https://www.pm.gov.au/media/national-royal-commission-black-summer-bushfires-established
40 Tolhurst, K 2020, 'We have already had countless bushfire inquiries. What good will it do to have another?', *ABC News*, 16 January, accessed on 26 July 2021, https://www.abc.net.au/news/2020-01-16/we-do-not-need-bushfire-royal-commission-this-is-why/11870824
41 Marshall, P 2020, 'The last thing we need is another bushfire royal commission', *Sydney Morning Herald*, 19 January 2020, accessed on 25 July 2021, https://www.smh.com.au/

national/the-last-thing-we-need-is-another-bushfire-royal-commission-20200117-p53sev.html

42 Marshall 2020.
43 Binskin 2020, p. 345.
44 Binskin 2020, p. 357.
45 Watson, J & Docherty, S 2020, 'Australia's bushfire royal commission follows a string of others. How do we measure their worth?', *ABC Radio National*, 20 February, accessed on 25 July 2021, https://www.abc.net.au/news/2020-02-20/australia-bushfires-royal-commission-what-is-their-real-value/11978286
46 Tonacia 2020, p. 5.
47 Tonacia, K 2020 'Picking up the Peaces: A human centred, community focused approach to disaster management', *Royal Commission into National Natural Disaster Arrangements Submissions, April 2020*, Document No. NND.600.00279, p. 5.
48 Rose, D 2020, 'Exhibit 3–008.001 video evidence of Denis Rose', *Royal Commission into National Natural Disaster Arrangements Submissions*, Document No. RCN.500.001.2351.
49 Binskin 2020, p. 43.
50 Binskin 2020, p. 21.
51 Rose 2020.
52 Williamson, B, Markham, F & Weir, J 2020, *Aboriginal Peoples and the response to the 2019–2020 bushfires*, Working Paper No. 134/2020, Centre for Aboriginal Economic Policy Research, Australian National University, p. iii.
53 Cunsolo, A & Ellis, NR 2018, 'Ecological grief as a mental health response to climate change-related loss', *Nature Climate Change*, Vol. 8, No. 4, pp. 275–81.
54 Amy Thunig as cited in Williamson, Markham & Weir 2020, p. 12.
55 Lorena Allem as cited in Williamson, Markham & Weir 2020, p. 12.
56 Vanessa Cavanagh cited in Williamson, Markham & Weir 2020, p. 12.
57 Victor Steffensen in *Q+A Bushfires Special* 2020, television program, *ABC*, 3 February, accessed on 18 October 2021, https://www.abc.net.au/qanda/2020-03-02/11906192
58 Noel Webster as cited in Williamson, Markham & Weir 2020, p. 13.
59 Noel Webster as cited in Knowles, R 2020, 'Talk is cheap, time for action: Yuin man promotes traditional fire management practices', *National Indigenous Times*, 20 January, accessed on 14 September 2021, https://nit.com.au/talk-is-cheap-time-for-action-yuin-man-promotes-traditional-fire-management-practices/
60 Jenny Brown as cited in Australian Red Cross 2021, *Australian Bushfires Report January-December 2020*, accessed 15 January 2021, https://www.redcross.org.au/bushfirereport/ p. 2.
61 Sherrie as cited in Australian Red Cross 2021, p. 13.
62 Sherrie as cited in Australian Red Cross 2021, p. 13.
63 Sherrie as cited in Australian Red Cross 2021, p. 13.
64 Foley, M 2022, 'Indigenous rangers program doubles with $636 million boost', *Sydney Morning Herald*, 29 March, accessed on 2 April 2022, https://www.smh.com.au/politics/federal/indigenous-rangers-program-doubles-with-636-million-boost-20220323-p5a79w.html
65 See Goodman & Horton 2021, Copernicus 2022 and Yeranian 2021.
66 Australia ICOMOS & Blue Shield 2020, 'Rapid assessment – bushfire damage to heritage places', Australia ICOMOS Bushfire Response Resources, accessed 2 April 2020, https://australia.icomos.org/resources/australia-icomos-heritage-toolkit/bushfire-response-resources/
67 Climate Council 2020, *Communication Guide: Extreme Weather*, accessed 2 April 2022, https://www.climatecouncil.org.au/resources/extreme-weather-communication-guides/
68 Climate Council 2020.
69 Climate Council 2020.

References

ABC News 2009, '"Worst ever" fire risk threatens Vic', 6 February, accessed on 18 October 2021, https://www.abc.net.au/news/2009-02-06/worst-ever-fire-risk-threatens-vic/285892

Albrecht, G 2005, 'Solastalgia: a new concept in human health and identity', *PAN (Philosophy, Activism, Nature)*, Vol. 3, pp. 44–59, doi:10.4225/03/584f410704696

Albrecht, G 2012, 'The age of solastalgia', *The Conversation*, 7 August, accessed on 17 May 2021, https://theconversation.com/the-age-of-solastalgia-8337

Australia ICOMOS & Blue Shield 2020, 'Rapid assessment – bushfire damage to heritage places', Australia ICOMOS Bushfire Response Resources, accessed on 2 April 2020, https://australia.icomos.org/resources/australia-icomos-heritage-toolkit/bushfire-response-resources/

Australian Red Cross 2021, *Australian Bushfires Report January-December 2020*, accessed on 15 January 2021, https://www.redcross.org.au/bushfirereport/

Binskin, M 2020, *Royal Commission into National Natural Disaster Arrangements Report*, Commonwealth of Australia, accessed on 15 November 2020, https://naturaldisaster.royalcommission.gov.au/publications/royal-commission-national-natural-disaster-arrangements-report

Block, K, Molyneaux, R, Gibbs, L, Alkemade, N, Baker, E, MacDougall, C, Ireton, G & Forbes, D 2019, 'The role of the natural environment in disaster recovery: "we live here because we love the bush"', *Health & Place*, Vol. 57, pp. 61–69, doi:10.1016/j.healthplace.2019.03.007

Bureau of Meteorology (BOM) 2009, 'Special climate statement 14', Australian Government, http://www.bom.gov.au/climate/current/statements/scs14.pdf

BOM 2019a, 'Remembering Black Saturday: the extraordinary weather behind Victoria's 2009 bushfires', *BOM Blog*, blog post, accessed on 26 February 2021, https://media.bom.gov.au/social/blog/2025/remembering-black-saturday-the-extraordinary-weather-behind-victorias-2009-bushfires/

BOM 2019b, *Tracking Australia's Climate Through 2019*, accessed on 1 July 2021, http://www.bom.gov.au/climate/updates/articles/a036.shtml

Burgess, T, Burgman, JR, Hall, S, Holmes, D & Turner, E 2020, *Black Summer: Australian Newspaper Reporting on the Nation's Worst Bushfire Season*, Monash Climate Change Communication Research Hub, Monash University, Melbourne.

Brown, D 2014, 'Preservation in the 21st century: preservation is a political movement', *National Trust for Historic Preservation*, accessed on 8 April 2021, https://forum.savingplaces.org/blogs/forum-online/2014/07/16/preservation-in-the-21st-century-preservation-is-a-political-movement

Carlowicz, M & Schollaert Uz, S 2017, 'El Niño Pacific wind and current changes bring warm, wild weather', *NASA Earth Observatory*, 14 February, accessed on 2 April 2022, https://earthobservatory.nasa.gov/features/ElNino

Climate Council 2020, *Communication Guide: Extreme Weather*, accessed 2 April 2022, https://www.climatecouncil.org.au/resources/extreme-weather-communication-guides/

Copernicus 2022, 'South America sees record wildfire activity in early 2022', *Copernicus Atmospheric Monitoring Service*, 2 March, accessed on 2 April 2022, https://atmosphere.copernicus.eu/south-america-sees-record-wildfire-activity-early-2022

Cunsolo, A & Ellis, NR 2018, 'Ecological grief as a mental health response to climate change-related loss', *Nature Climate Change*, Vol. 8, No. 4, pp. 275–281, doi:10.1038/s41558-018-0092-2.

Dale-Hallett, L 2009, 'Victorian bushfires collection', *Museums Victoria*, accessed on 29 October 2021, http://collections.museumsvictoria.com.au/articles/3032

Ellem, A 2011, 'In support of public and private treasures: the AICCM Victorian Division's disaster response to the Victorian fires of 2009', *AICCM Bulletin*, Vol. 32, No. 1, pp. 190–195.

Foley, M 2022, 'Indigenous rangers program doubles with $636 million boost', *Sydney Morning Herald*, 29 March, accessed on 2 April 2022, https://www.smh.com.au/politics/federal/indigenous-rangers-program-doubles-with-636-million-boost-20220323-p5a79w.html

Garnaut, R 2008, *The Garnaut Climate Change Review Final Report*, Cambridge University Press, Port Melbourne, Australia.

Gibbs, L, Molyneaux, R, Harms, L, Gallagher, C, Block, K, Richardson, J, Brandenburg, V, O'Donnell M, Kellett, C, Quinn, P, Kosta, L, Brady, K, Ireton, G, MacDougall, C & Bryant, R 2020, *10 Years Beyond Bushfires Report*, University of Melbourne, Melbourne, Australia.

Gillard, J 2009, *Condolence Speech - Victorian Fires*, Ministers' Media Centre Dept. of Education, Skills and Employment, accessed on 14 May 2021, https://ministers.dese.gov.au/gillard/condolence-speech-victorian-fires

Goodman, J & Horton, J 2021, 'Wildfires: how are they linked to climate change?', *BBC Reality Check*, 11 August, accessed on 2 April 2022, https://www.bbc.com/news/58159451

Heberger, M 2012, 'Australia's millennium drought: impacts and responses' in PH Gleick (ed) *The World's Water, Volume* 7, Island Press, Washington, DC, pp. 97–125.

Hughes, L, Steffen, W, Mullens, G, Dean, A, Weisbrot, E & Rice, M 2020, *Summer of Crisis*, Climate Council of Australia, https://www.climatecouncil.org.au/resources/summer-of-crisis/

Knowles, R 2020, 'Talk is cheap, time for action: Yuin man promotes traditional fire management practices', *National Indigenous Times*, 20 January, accessed on 14 September 2021, https://nit.com.au/talk-is-cheap-time-for-action-yuin-man-promotes-traditional-fire-management-practices/

Marshall, P 2020, 'The last thing we need is another bushfire royal commission', *Sydney Morning Herald*, 19 January, accessed on 25 July 2021, https://www.smh.com.au/national/the-last-thing-we-need-is-another-bushfire-royal-commission-20200117-p53sev.html

Moncrief, M 2009, '"Worst day in history": Brumby warns of fire danger', *Sydney Morning Herald*, 6 February, accessed on 24 February 2021, https://www.smh.com.au/national/worst-day-in-history-brumby-warns-of-fire-danger-20090206-7zf1.html

Morrison, S 2020, 'National Royal Commission into Black Summer bushfires established' [press release], Prime Minister of Australia, accessed on 17 July 2021, https://www.pm.gov.au/media/national-royal-commission-black-summer-bushfires-established

O'Gorman, D 2020, 'Forward' in World Wildlife Fund for Nature Australia 2020, *Australia's 2019–20202 Bushfires: The wildlife toll*, p. 1, https://www.wwf.org.au/news/news/2020/3-billion-animals-impacted-by-australia-bushfire-crisis#gs.e6ic8d

O'Neill, SJ & Handmer, J 2012, 'Responding to bushfire risk: the need for transformative adaptation', *Environmental Research Letters*, Vol. 7, pp. 1–7, doi:10.1088/1748-9326/7/1/014018.

Osborne, M 2022, 'The Western USA is experiencing the worst drought in more than 1,200 years', *Smithsonian Magazine*, 17 February, accessed on 2 April 2022, https://www.smithsonianmag.com/smart-news/the-western-us-is-experiencing-the-worst-megadrought-in-more-than-1200-years-180979590/

Q+A Bushfires Special 2020, Television program, ABC, 3 February, accessed on 18 October 2021, https://www.abc.net.au/qanda/2020-03-02/11906192

Rapport, D & Whitford, WG 1999, 'How ecosystems respond to stress', *Bioscience*, Vol. 49, No. 3, pp. 193–203, doi:10.2307/1313509

Read, P 1996, *Returning to Nothing: The Meaning of Lost Places*, Cambridge University Press, Cambridge.

Rose, D 2020, 'Exhibit 3-008.001 video evidence of Denis Rose', *Royal Commission into National Natural Disaster Arrangements Submissions*, Document No. RCN.500.001.2351.

Tolhurst, K 2020, 'We have already had countless bushfire inquiries. What good will it do to have another?', *ABC News*, 16 January, accessed on 26 July 2021, https://www.abc.net.au/news/2020-01-16/we-do-not-need-bushfire-royal-commission-this-is-why/11870824

Tonacia, K 2020, 'Picking up the peaces: a human centred, community focused approach to disaster management', *Royal Commission into National Natural Disaster Arrangements Submissions*, Document No. NND.600.00279.

Wang, B, Luo, X, Yang, Y, Sun, W, Cane, MA, Cai, W, Yeh, S & Liu, J 2019, 'Historical change of El Niño properties sheds light on future changes of extreme El Niño', *Proceedings of the National Academy of Sciences of the United States of America*, Vol. 116, No. 45, pp. 22 512–22 517, doi:10.1073/pnas.1911130116

Watson, J & Docherty, S 2020, 'Australia's bushfire royal commission follows a string of others: How do we measure their worth?', *ABC Radio National*, 20 February, accessed on 25 July 2021, https://www.abc.net.au/news/2020-02-20/australia-bushfires-royal-commission-what-is-their-real-value/11978286

Webb, C 2009, 'Marysville's past is pouring back in', Sydney Morning Herald, August 7, accessed on 29 October 2021, https://www.smh.com.au/national/marysvilles-past-is-pouring-back-in-20090806-ebk2.html

Webb, C 2010, 'Marysville's history rises from the ashes to find new home', *The Age*, 7 June, accessed on 29 October 2021, https://www.theage.com.au/national/victoria/marysvilles-history-rises-from-ashes-to-find-new-home-20100606-xna7.html

Williamson, B, Markham, F & Weir, J 2020, *Aboriginal Peoples and the Response to the 2019–2020 Bushfires*, Working Paper No. 134/2020, Centre for Aboriginal Economic Policy Research, Australian National University.

Yeranian, E 2021, 'North Africa fires spread to Libya and Tunisia', *VOA News*, 13 August, accessed on 2 April 2022, https://www.voanews.com/a/middle-east_north-africa-fires-spread-libya-and-tunisia/6209549.html

7

THE LOCAL IN THE GLOBAL: COMMUNITY IMPACT AND RESPONSES

Robyn Sloggett and Marcelle Scott

This book is underwritten by two premises: that access to heritage contributes to how people can prepare, respond and recover from the impacts of severe natural disasters; and that heritage gives agency and meaning to people and communities as they recover from the catastrophic effects of climate change. Previous chapters have examined the experiences of slow onset (changes in temperature, seasons and weather) and rapid onset natural disasters (bushfires, floods, earthquakes and tsunamis) in three distinct populations (bush fire affected communities in Australia, East Timorese diaspora and Indigenous communities). While some commentators have argued for clear delineation of natural disasters from climate change-induced disasters, geologists are identifying a causal confluence between the two.[1] Despite differences in emphasis amongst some researchers, for practical purposes such as disaster mitigation, recovery or adaptation, the cascading effects of natural disasters and climate crisis-induced disasters on communities have the same impact, compounding existing inequity, vulnerability and dependence. Because this convergence of risks needs to be understood in any attempts to preserve, protect and ensure continuing access to heritage in all its forms, this chapter explores the links between climate change-induced emergencies, natural hazard disasters, and community responses to climate change.

The climate crisis is not a social leveller. In the 2018 *Summary for Policymakers*, the IPCC reported 'high confidence' that the climate crisis will increase disadvantage by 'disproportionately' affecting already disadvantaged and vulnerable populations.[2] For this reason, in the 2014 IPCC AR5 report, the chapter 'Small Islands' emphasised support for locally led programs, rather than externally driven projects, by 'empowering communities and optimizing the benefits of local practices that have proven to be efficacious through time'.[3] The need for locally led and properly contextualised decision-making has been a theme throughout this book. In 2019, the International Council on Museums and Sites (ICOMOS)

DOI: 10.4324/9781003163312-8

published a position paper that examined how cultural heritage can contribute to climate actions and solutions. The paper discussed the lack of professional engagement with policy development and the importance of value-based approaches and people-centred participatory governance. The authors also called for climate change to 'become a baseline competency of heritage management' emphasising that 'Cultural heritage is both impacted by climate change and a source of resilience for communities'.[4]

With this requirement for local empowerment and agency and the need for action to occur at a global level, this chapter examines the tension inherent in the phrase 'Think global, act local', and explores ways in which this dichotomy is manifested and responded to at a local level.

Small community collecting organisations

Indigenous art centres

In the twentieth century, Indigenous communities across the globe developed community-based cultural centres, keeping places and museums to hold, protect, preserve, provide access to, display and manage their cultural heritage. In Australia, hundreds of Indigenous knowledge centres, keeping places and art centres hold, protect and develop locally significant cultural material. Many are in very remote areas that are vulnerable socially, economically and, increasingly due to the climate crisis, environmentally. In 2007 the Australian Federal Government's Inquiry into the Indigenous art and craft sector found the sector was estimated to be worth between $400–$500 million to the Australian economy.[5] The report also found that Indigenous art centres have been important in the advancement of Aboriginal women, supporting 'personal development and self-esteem, financial independence and empowerment within their communities'.[6]

In 2017, the Arnhem, Northern and Kimberley Artists Aboriginal Corporation (ANKA). which is the peak body for Aboriginal artists and art centres across the Northern Territory and the Kimberley region of Western Australia, commissioned a study of the risk to cultural material held in its members' art centres. In the foreword to the report, Djambawa Marawili, ANKA's Chair and senior leader of the Madarrpa clan of the Yolŋu people of northeast Arnhem Land in northern Australia, argues for the critical importance of local Indigenous art-centres. He links the protection and conservation of cultural material with community well-being, employment and education, arguing that:

> Safe keeping makes Aboriginal people feel really strong and proud and alive. Caring for collections is a really significant role. It is important to give real jobs to Aboriginal people caring for and looking after objects in their community museums. That way they can also learn to share the knowledge and the patterns and the stories for future generations ... To keep culture alive in our blood and our soul, we need to continue looking after and

> caring for these collections. To have other tools to teach so that art and patterns will be safe, clean and well looked after. This is important; it is the history of Australia.[7]

Marawili identifies other reasons why collections held locally present exigent demands for preservation, arguing that they have value in teaching culture to 'the lost ones who don't know who they are' and are necessary 'to keep our sacred objects alive ... to strengthen culture and strengthen ourselves'.[8] He argues that as knowledge is embedded in culture then strong culture keeps knowledge alive. With the role of Indigenous Traditional Knowledge forming an important part of UNESCO's approach to the climate crisis, through its LINKS (Local and Indigenous Knowledge Systems) programs, the role of Indigenous art centres, and associated organisations is likely to become even more important.[9]

The need to preserve culture locally, and within communities, also supports the survival of the complex ontological networks and traceries of knowledge, identity, authority and decision-making that provide communities with strength, resilience and choice when faced with overwhelming threats, such as colonisation or climate change. In the debates on climate change and in the formulation of effective responses to climate change, the idea of continuity, and particularly of cultural continuity, is often submerged by discussions of sustainability, resilience, adaptation and transformation. Marawili, however, makes a strong case for considering cultural continuity, rather than adaption, as the key philosophical concept to be prioritised in the face of threats to communities. Arguments for adaptation are multi-faceted, and often predicated on the assumption that continuity will be sacrificed for survival. Marawili argues for the need to emphasise the value of cultural practice that can continue to link to country.

Small historic societies and community organisations

Around the world, small historical societies, local archives, and community groups hold records that reflect the identity of their local community. Fiona Cosson, in exploring the role of local history societies in the United Kingdom, asserts that:

> Personal papers and narratives enable the researcher to move away from state and official records and narratives, from the manor house to the terraced house, from the parsonage to the factory. As Raphael Samuel suggested, with these types of sources, 'the historian can draw up fresh maps, in which people are as prominent as places'.[10]

It is within this role of reassembling history that the heroic narrative of local heritage can be effectively understood.

With this in mind, a survey of 13 historical societies and heritage community organisations across southeast Australia was completed in order to obtain a snapshot of how local Australian collecting organisations might think about climate change, and consequently, to understand what leadership role such organisations may play in local climate change debates. The survey was conducted after the devastating 2021 bushfires and 2020 floods, and representatives of these groups were invited to respond to six online survey questions:

1. Are you aware of people in your community talking about climate change?
2. Are you aware of members of your organisation or community group concerned about the effects of climate change?
3. Are there any specific events that have focused people's attention on climate change?
4. Does your organisation or community group hold archives or material that is relevant to research on climate change?
5. Are members of your organisation or community group working with others (politicians, educators, health workers, museums, or any others) on the issue of climate change?
6. What do you think are the biggest threats to your community from climate change?
7. Do you have any other comments or ideas you would like to share?[11]

The results indicated a general consensus that climate change was an area of interest amongst members of the respective organisations, with only one respondent reporting that they were not aware of climate change being talked about in their group. Another noted that while people were not talking about climate change, nevertheless, the bushfires had focused the attention of their members on climate change. Overwhelmingly, and unsurprisingly, the majority of respondents reported that it was the recent bushfire events that was most clearly focusing their members' thoughts on climate change; while those affected by flood also noted floods as formative. One respondent from Yackandandah Historical Society identified fires in 2003, 2006, 2011 and 2013 as being significant, another from Tathra Wharf Museum identified bushfires in 2018, and nine noted the impact of the 2019–2020 Black Summer bushfires on members' awareness of climate change. Only three respondents, all representing coastal communities, identified floods as influencing opinions about climate change amongst their membership. These floods had all occurred locally, in 2007 (Painsville), 2012 (Nambucca Heads), and 2020–2021 (Port Macquarie). Nambucca Heads also noted the 1974 river mouth retaining wall failure.

When asked to consider if their organisation held archival or other material relevant to climate change research, only four thought this was the case. They identified this as: 'diaries of local people/families'; 'anecdotal evidence recorded via interview or social media, Newspaper reports, Photographs'; 'The NSW Bushfires 2019/2020' and 'Archival Newspapers, Personal archival accounts of bushfires and other natural disasters, Historic photographs and maps'; and 'Newspaper clippings and reports on flooding drought and major events over the past 100 years'. As discussed in Chapter 2, new research methods such as citizen science projects are enabling vast amounts of locally-held records to be digitised for use in climate change research across the globe. The amount of material that may exist, but is unidentified as relevant to climate change research, points to the need for more work to be done in ensuring that valuable data for climate change research is identified, preserved and made accessible for future research.

Five points emerge from the responses to the 2021 survey. First, there is a perceived alignment between an interest in community heritage and an interest in climate change with respondents reflecting on the significance of local rather than global events. Second, while the focus was on local impacts, these were recognised as the result of global climate change. Third, although some organisations identified information and historic records relevant to climate change in their collections, a number did not. Of those who did not, it is likely that they hold this kind of material but that it has not been identified as having relevance. This means that there is the opportunity for extended research programs to identify such material and, at the same time, raise awareness of the value of historical material for climate change research at a local level. Fourth, at a local level, the increasing cost of responding to climate is recognised as a problem. Finally, there appears to be a strong group of community members who are committed to their community organisation and are also interested in the issues raised by climate change, including issues relating to losses in the natural environment.

Built heritage

The Philippines: Typhoon Haiyan/Yolanda of 2013

The Philippines is an archipelago of around 7,500 islands. The Filipino population of around 111 million people inhabit some 2,000 of these islands. This is a part of the world that is particularly susceptible to natural disasters including earthquakes, volcanic eruptions, tropical cyclones, floods and landslides. Just over half of the population live in rural areas, where there is a high reliance on local infrastructure and multiple forms of transport for food production and supply. This makes the country highly vulnerable to adverse impacts from natural disasters.

The Philippines is home to significant diversity in beliefs, customs, language, and culture. In the Central Visayan region in Bohol, a large island located in the centre of the Philippines, are some of the country's most significant churches. Representing the global authority of the Church, they are testaments to colonisation. Embodying local beliefs and customs, and acting as foci for ceremonies, community activities, festivals, and personal rituals around births, deaths, marriages and other significant events, the churches of Bohol are also a demonstration of Filipinos' responses to colonisation. They also hold significant material culture in the form of artwork, archives, ecclesiastical fittings and fixtures, and textiles, and are important repositories of knowledge about Filipino beliefs and customs. Church cultural records also preserve intangible practices. Liturgical songbooks document the sounds heard in the Church; embroidered liturgical robes demonstrate the crafts associated with the Church; Church archives document the daily lives of the people who provided goods and services to the Church; and the registers of births deaths and marriages record the ebbs and flows and the joys and tragedies of the population (Fig. 7.1).

In the Visayas region, on 15 October 2013, an earthquake measuring 7.2 on the Richter Scale occurred on Bohol. The earthquake caused widescale destruction, affecting over 1.2 million people, killing 222, injuring 976 and displacing over 340,000. Around 14,5000 buildings were destroyed and over 79,000 were damaged.[12] The earthquakes damaged 25 churches across Bohol and completely destroyed St. Vincent Ferrer Church in Maribojoc and Our Lady of Light Church in Loon. The clocktower and facade of San Pedro Apostol Parish Church in Loboc (commonly known as the Loboc Church) were severely damaged (Figs. 7.2–7.4).

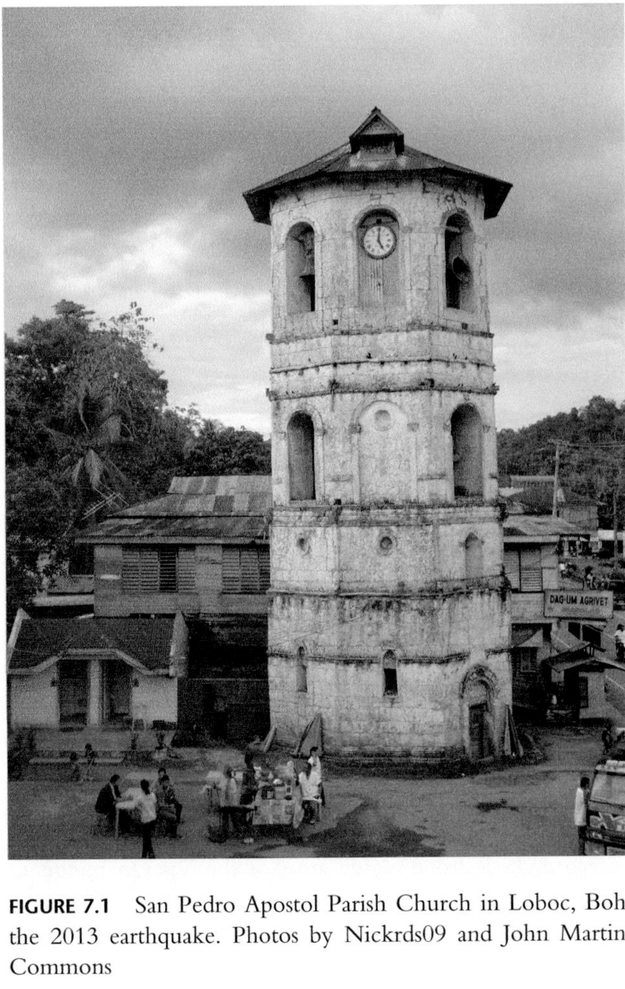

FIGURE 7.1 San Pedro Apostol Parish Church in Loboc, Bohol. Clocktower before the 2013 earthquake. Photos by Nickrds09 and John Martin Perry via Wikimedia Commons

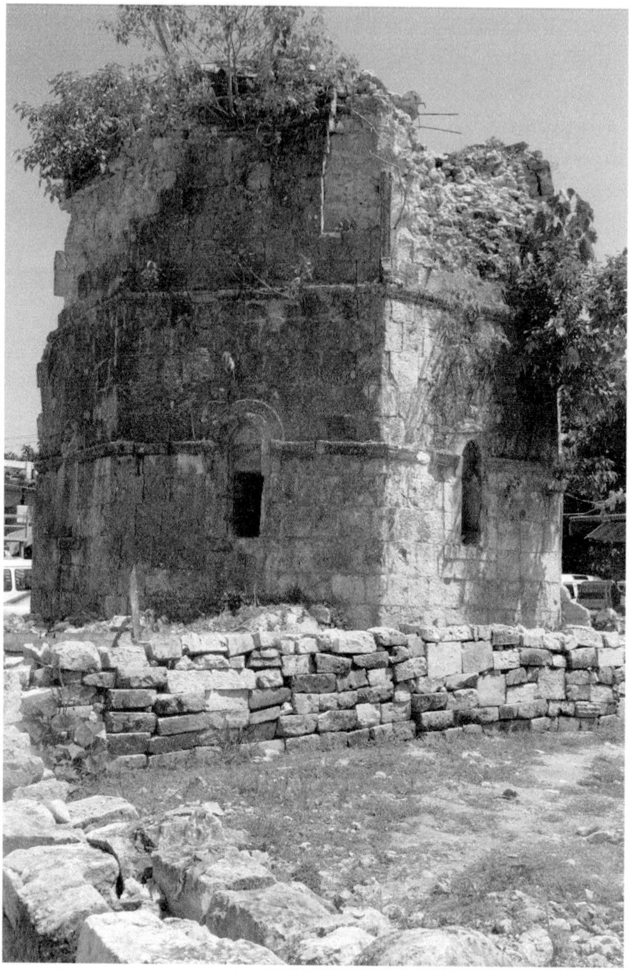

FIGURE 7.2 San Pedro Apostol Parish Church in Loboc, Bohol. Clocktower after the 2013 earthquake. Photos by Nickrds09 and John Martin Perry via Wikimedia Commons

The damage in Bohol was extensive and catastrophic, and emergency service rushed to the island to begin triage relief and salvage operations. Three weeks later, in early November 2013, Typhoon Haiyan (locally known as Yolanda) began to develop in the western Pacific, just north of Micronesia. Typhoon Haiyan soon registered as a Category 5 'super typhoon', making it one of the most powerful tropical storms in living memory, and on 8 November Typhoon Haiyan hit the east coast of The Philippines lashing communities with winds that reached 315 km/h, the highest on record. It destroyed large areas of the Visayan region in central Philippines, home to around 18 million people. Around 6,000 people were killed and 1,800 were registered missing. More than one million houses were destroyed. On the island of Leyte, 90% of Tacloban City was destroyed.

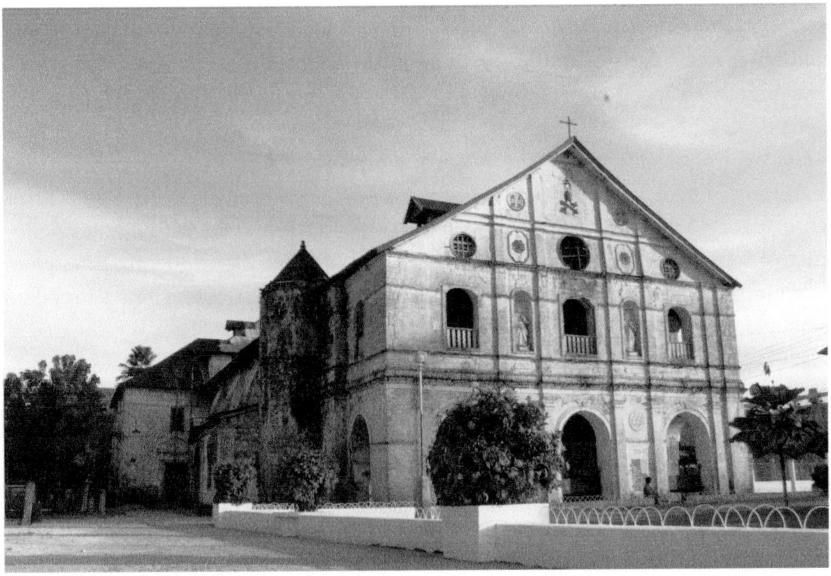

FIGURE 7.3 Loboc Church façade before the 2013 earthquake. Photos by SieNash and John Martin Perry via Wikimedia Commons

Loboc Church façade before the 2013 earthquake.

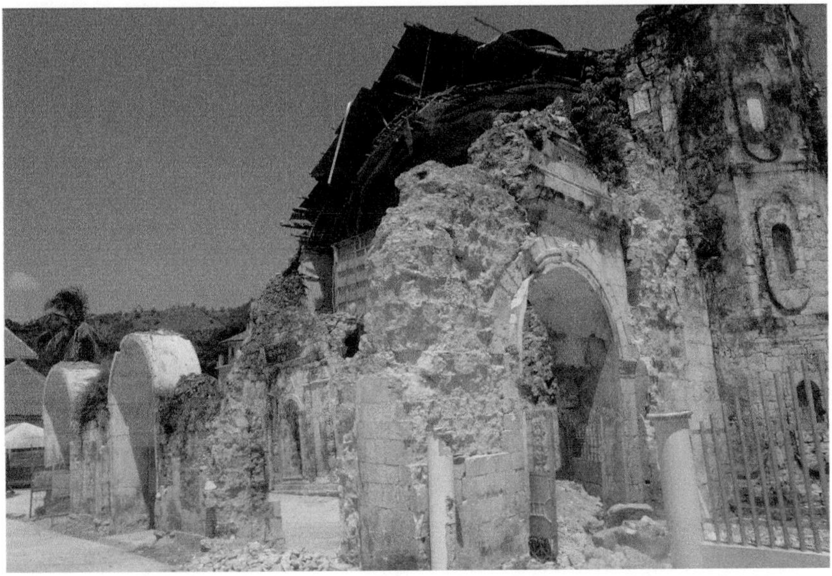

FIGURE 7.4 Loboc Church façade after the 2013 earthquake. Photos by SieNash and John Martin Perry via Wikimedia Commons

Loboc Church façade after the 2013 earthquake.

Immediately, resources that were being directed to Bohol were redirected to the typhoon-ravaged parts of the Visayas and other areas of need in The Philippines. This directly reduced the support that was available, and therefore the rate of recovery in Bohol.

Cyclonic activity is one of the major hazards in the Asia-Pacific region and increased major cyclonic activity is an outcome of climate change. In 2015 and 2016, two category 5 storms, the highest on the classification scale with sustained destructive winds that bring gusts on open flat land of more than 280 kph, devastated communities in Vanuatu and in Fiji. In Vanuatu in March 2015, Cyclone Pam killed 16 people, over 90% of buildings were destroyed and total damage was estimated at US$600m. In Fiji in February 2016 Cyclone Winston brought winds of over 306 kph, killing 44 people with total damage estimated at around US$1.4 billion. It was identified as the strongest and costliest cyclone in South Pacific history.

Examples such as these demonstrate that, in the face of such catastrophic disasters, heritage professionals are unlikely to be able to respond, and local heritage experts will be in the front line. For this reason, investing in and building programs to support local heritage experts before a disaster eventuates, is the most effective way for the broader heritage sector to ensure heritage is protected, preserved and cared for during and after catastrophic disasters. Providing resources, education, and improved financial capacity, and supporting local agency in decision-making, are practical outcomes of understanding that vulnerable communities are likely to become more vulnerable as cascading threats reduce capacity and divert valuable finite resources.

Cultural practices

The Pacific drought: 2015–2017

The fluctuating cycles of El Niño and La Niña, discussed in the context of unprecedented bushfires in Chapter 6, are also responsible for an increase in the severe droughts that occur across Asia, the Pacific and the Americas.[13] Drought is a sneaking invader that builds over time and is recognised by the damage it causes. As such, droughts are slow-onset hazards often lasting years rather than days, and their effects can be catastrophic. Small Island nations are particularly vulnerable with low levels of natural groundwater, and a high reliance on rainfall. In a paper that examined drought and its effects in Pacific Island Countries and Territories (PICTs) Iese et al explained that droughts affecting PICTS are poorly understood because

> the cause-effect relationships between drought and the variety of factors that contribute to drought are complicated and not satisfactorily monitored, modelled or predicted.[14]

They also note that, for Pacific islands and atolls in particular, information is lacking 'on the direct and indirect impacts of the different types of drought and how these vary across locations, communities and industries'.[15] For communities that rely on oral and performative records rather than written records, evidence of slow-impact disasters are difficult to find. Iese et al note that anecdotal accounts exist of pre-colonial droughts that caused famines across the Pacific and that early European settler records provided written accounts but that overall 'long-term, objective documentation of droughts in PICTs is limited'.[16] Complementing literature surveys with narrative analysis Iese et al identified a range of impacts of past drought in the Pacific, including severe droughts in Samoa in 1972, 1983, 1987 and 1997 that 'triggered forest fires and destroyed 80% of food'.[17] They also noted increases in pests and diseases that affected crops, and a loss of economic crops and livestock deaths, along with examples of population relocation. Iese et al also emphasised the difficulty many communities had in recovering from drought due to the cyclic nature of extremes in typhoon activity and drought, with risks now exacerbated by rising sea levels.

Droughts affect cultural heritage in a number of ways, disrupting cultural celebrations, interfering with traditions relating to kinship responsibilities, impacting the preparation of traditional foods, and creating a reliance on new economies. During the 2015–2016 Pacific drought in Vanuatu around 30% of the population relied on food aid. Villagers reported

> difficulties in the preparation of traditional dishes such as *laplap* (grated root vegetables wrapped in a specific leaf and baked in a stone oven) … and *nalot* (roasted breadfruit pounded into a gelatinous mass, covered with coconut milk and eaten communally), due to loss of key ingredients, *laplap* leaf and breadfruit, after TC Pam and the El Niño event.[18]

This, in turn, led to a loss of traditional knowledge. In Vanuatu, women noted that methods for food preparation were no longer passed down from mother to daughter, as a result people did not spend time together in food preparation or in sharing food communally, activities that are done 'all while talking, laughing and sharing stories'.[19] Although not noted in the study, this in turn, is likely to have affected the manufacture of traditional cooking implements, weaving of food containers and objects for food presentation, carved spoons, and so on that is also part of the cycle of the retention of traditional skills. Droughts also have an impact on the types of plant and animal material that is available for the production of traditional handicrafts and ceremonial items.

In rural Papua New Guinea the need for emergency food assistance 'exacerbated vulnerability drivers, such as reduced self-reliance and adaptive capacity'.[20] Social disruption occurred as the death rate rose with a reported rise in alcohol consumption. In Tuvalu people became sick from drinking poor quality water, traditional crops died resulting in the 'abandonment of pulaka cultivation and associated loss of traditional knowledge' and invasive seaweed species increased. Leadership and political stability were also challenged as:

> Tuvaluans also consider droughts as a punishment from God for sin or bad traditional leadership. Some islands replaced their chiefs because of the negative impacts of droughts.[21]

Sustainability, resilience, adaptation and transformation

The causal confluence between climate change and natural disasters is an area of increasing interest to geophysicists, geomorphologists, and geologists. Research indicates that impacts from climate change, such as reduced atmospheric pressure, are directly impacting aspects of geomorphology that, for example, link earthquakes and typhoons. This was explored over a decade ago by Liu, Linde and Sacks who modelled the relationship between low pressure, typhoons and slow earthquakes.[22] Bill McGuire, Emeritus Professor of Geophysical & Climate Hazards at University College London, wrote extensively about these links in his book *Waking the Giant: How Changing Climate Triggers Earthquakes, Tsunamis and Volcanoes*.[23] More recently structural geologist Jane Cunneen identified research that links climate change to sea level rises, landslides, ice shelf collapse, iceberg calving, volcanic activity and earthquakes.[24] Other researchers identify the links between climate change and 'third order' events such as infectious diseases, increased injury and mortality, and food and economic insecurity.[25] For heritage experts the cause of the threat is, generally, of less interest than the conjoined effects of these threats. As the 2018 IPCC Special Report *Global Warming of 1.5°C* published in its 'Summary for Policymakers':

> Climate-related risks for natural and human systems are higher for global warming of 1.5°C than at present, but lower than at 2°C (*high confidence*). These risks depend on the magnitude and rate of warming, geographic location, levels of development and vulnerability, and on the choices and implementation of adaptation and mitigation options (*high confidence*).[26]

The 'Summary' also made the point that:

> There are limits to adaptation and adaptive capacity for some human and natural systems at global warming of 1.5°C, with associated losses (*medium confidence*). The number and availability of adaptation options vary by sector (*medium confidence*).[27]

For this reason, it makes sense to consider where increased risks from climate change sit within sets of cascading risks that are likely to push communities beyond their adaptation and adaptive capacity. As heritage remains a relatively low priority in intergovernmental reports on climate change then understanding at what point and in what way to support communities becomes a critical decision for heritage experts.

In their paper 'Sustainability, resilience, adaptation, and transformation: tensions and plural approaches', Zanotti et al explore the complexity of use and the ubiquity of these four terms.[28] They note that although they are considered as interrelated terms, they have distinct meanings and reflect distinct paradigms with different origins. The authors also note that the term 'sustainability' has become the dominant term 'to be prioritized as goals or outcomes in local to global environmental governance and management' and developed as frameworks for other complex systems such as environmental stewardship, economic development, political organisation, and equity concerns, as well as reflecting intra- and inter-generational justice.[29] They note that the term 'sustainability' has a well-acknowledged normative dimension but that this is not the case with 'resilience', 'adaptation' and 'transformation'. This, they conclude, 'has implications for equitable outcomes'.[30] They emphasise that:

> it is important for any resilience-, adaptation-, or transformation-focused policy, program, or intervention to ask questions such as who the targets are, how the processes occur, and who bears the consequences as results of the policy, program, or intervention (Johnson et al. 2018). Asking these questions would make explicit the implicit normative dimensions of policies, programs, or interventions as well as their desired outcomes.[31]

Dr Ana Labrador, past Deputy Director, National Museum of The Philippines, pointed to problems with implicit normative assumptions explaining that

> resilience is something that is a bit controversial here because people would say 'oh the Filipinos can take on anything because they are so resilient' so a lot of us have kind of grown to detest that because that means that we can be abused … for the Filipinos really an important concept [is] that we are versatile … if things cannot happen we will find ways to address that … you know this resilience is like a mindset that keeps us from fighting back.[32]

Marianne Tissandier, head of Collections and Conservation, from Musée de Nouvelle Calédonie pointed to a broader context identifying the climate crisis as creating impacts that are akin to a new wave of colonialism as communities realign their lives with the brutal realities of climate change.[33]

Djambawa Marawili argued that cultural heritage held locally, and managed by community is a major contributor to individual and community strength. Both Marawili's and Tissandier's observations accord with those of the Iñupiat communities of the North Slope Borough (NSB) of the Beaufort and Chukchi seas and surrounding inland areas, who formed one of the case studies in Zanotti et al's research. These communities, while affected by and responding to the challenges of climate change,

> invited us to critically approach policy oriented or scholarly approaches to resilience and sustainability and to consider the following: (1) how previous practices, such as the IWC [International Whaling Commission] ruling,

> continue to undermine local well-being; and (2) the language of sustainability, adaptive capacity, and resilience are entrenched in Western understandings of solutions that might superficially describe responses to change but do not have the historical context and cosmological grounding to be locally relevant.[34]

As Labrador and Tissandier identified, words have weight depending on their context within the heritage, history, and current issues of a community. The Iñupiat communities of the North Slope Borough (NSB) advised that the paradigms of 'resilience and sustainability' and the 'social-ecological systems approach' needed to be critically rethought to better understand and accommodate the alignment between the impacts of climate-change mitigation decision-making, past historical injustices and the need to incorporate Iñupiat worldviews.[35] In this way, and for these reasons, concepts of sustainability, resilience, adaptation, and transformation, when actualised as plans or programs must consider the normative and local, as well as professional approaches and opinions. Citing work by Smit et al, Thomsen, Smith & Keys point out that critical examination of the term 'adaptation' with a more focused specificity of: '(i) adaptation to what; (ii) who or what adapts; and (iii) how does adaptation occur?' helps to identify the normative in abstracted discussions of adaptation.[36] Similar questions can be asked aboutthe use of the terms 'sustainable', 'resilience' and 'transformation'.

Zanotti et al reflect Marawili's words when they argue that effective responses to significant change must acknowledge

> sovereign and self-determined futures as orienting frames instead of "sustainable" or "resilient" ones, emphasizing cultural continuity, honoring different generations, and supporting family and community as preferred pathways forward.[37]

For this reason, proper, sustained and effective investment in local heritage experts needs to be a critical part of any global response by heritage professionals to climate change.

Building strong community engagement

The role of communities in climate change mitigation and adaptation is core to the mission of many international organisations. The Sendai Framework for Disaster Risk Reduction 2015–2030 presents United Nations member states with 'concrete actions to protect development gains from the risk of disaster'.[38] The Framework identifies community engagement and mobilisation as critical in any attempts to prevent or reduce risk and harm from disasters including climate change, noting:

> There has to be a broader and a more people-centred preventive approach to disaster risk. Disaster risk reduction practices need to be multi-hazard and multisectoral, inclusive and accessible in order to be efficient and effective.[39]

Building strong communities is also a function of the United Nations' Sustainable Development Goals. Yet, the word 'community' is broad, and engaging communities is a complex proposition as one seemingly homogenous community will include smaller, overlapping communities. These sub-communities may intersect with other communities that may have different ideas and different requirements. In small rural communities, groups such as the local Returned Services League, the local school climate strike action group, or the local Historical Society, all have an interest in preserving what we have now for the future, and are potential allies for heritage professionals who are engaged in climate change research, risk assessment or disaster risk reducation. Yet their understanding of climate change, their political views about climate change, their understanding of the role they have to play, and their belief in their ability to have agency, may all be very different. For this reason,

As both Djambawa Marawili and Fiona Cosson argue, cultural material and heritage held in communities are central to community well-being and help in understanding community values and beliefs. Section 47 of the Sendai Framework, 'Means of implementation' identifies the needs for (a) coordinated support, (b) environmentally sound technology, science and inclusive innovation, (c) the use of thematic platforms of cooperation, and (d) the incorporation of disaster risk reduction measures across all sectors.[40] This last recommendation, which talks about multilateral and bilateral assistance programmes lists 'poverty reduction, sustainable development, natural resource management, the environment, urban development and adaptation to climate change'.[41] Making the case for heritage in all its forms to be included here as a separate category, perhaps as 'heritage preservation' seems like a grounded, and important piece of work.

While the work of the United Nations and its agencies and instrumentalities is crucial in bringing coordinated responses to the local level, it is advocacy, education and awareness that will underwrite any attempts at effective disaster response and remediation. As Chapter 1 illustrated, the story that people hold on to is the story that will impact actions and decision-making. Without understanding the heritage values embedded within a community, programs that call for the inclusion of local voices in decision-making, the utilisation of value-based approaches, a focus on people-centred processes and other exhortations to make the local part of the global, may fail to take account of the context in which programs are being progressed. As a result, such programs are also likely to fail.

In 2017, UNESCO published the outcomes of the two-year project 'Towards climate change resilience: minimising loss and damage in Pacific SIDS communities', which sought to

> generate and share new knowledge and raise awareness on loss and damage caused by the adverse impacts of climate change, particularly in Pacific Small Island Developing States (SIDS).[42]

The project focused on tourism and agriculture as areas at risk from the impacts of climate change and explored these risks in five Pacific Island SIDS; Cook Islands, Fiji, Samoa, Solomon Islands, and Timor-Leste. Heritage was not identified in the report. Given that local heritage is often touted as the basis of tourism, this is a challenging omission for heritage professionals who work with local communities to preserve local heritage. If done correctly, grounded research on community heritage and climate change can be effectively aligned to the global focus of UNESCO, IPCC, the Sendai Framework, and others. In order to be done correctly, it will require strong interdisciplinary teams across heritage-related professions that are led by community heritage experts. If done effectively, such research presents a very practical contribution that community heritage experts and heritage professionals can make to fill a significant gap in knowledge about climate change response.

As the survey of 13 community historical societies in rural, southeast Australia demonstrated, their members predominately link climate change to recent catastrophic events. If heritage, in all its forms, is to be protected from the impacts of the climate change, then it will be important to ensure that people in communities who have responsibility for local heritage also have access to information about the broad range of impacts from climate change. The answer to the question: How best to engage communities in discussions of how to make heritage adaptable and resilient so that it can be protected, preserved and sustained in the face of the cascading effects of climate crises? is not readily available. Sustained and co-ordinated research is necessary to address this question. Areas of research could include, as a first step, scoping studies that identify the specific communities, sub-communities and community networks that need to be engaged in climate change. A coordinated understanding of significances of community-held heritage, as opposed to the significance of individual items, places or collections would also be useful. This could integrate with studies that seek to understand and map the vulnerabilities of the particular forms of cultural heritage that is important to these communities. There is also considerable associated work to be done in assessing and mapping the risk of exposure to climate change to this cultural heritage, as well as assessing and mapping the hazards that are creating this risk. Finally, understanding the tolerance of a community for adaptive change, the levels of resilience within a community that result from access to community heritage, and the ways in which local heritage holds information that can assist in strategies for climate change mitigation, are all areas in which only very limited evidence-based data is currently available.

This chapter has shown that a coordinated program of research that fills the gaps that are so obvious in many of the reports of international organisations is an important advocacy tool for heritage professionals, but it is also an acknowledgement of professional responsibility in a time of global crisis. On a similar theme, the following chapter looks at the multiple contributions peak bodies and cultural organisations are making as they each respond in different ways to the heritage for which they are responsible.

Notes

1 See, for example, the work of Bill McGuire, Jane Cunneen and others discussed in more detail later in this chapter.
2 IPCC 2018, 'Summary for policymakers' in IPCC 2018, *Global Warming of 1.5°C. An IPCC Special Report on the Impact of Global Warming of 1.5°C above Pre-industrial Levels and Related Global Greenhouse Gas Emission Pathways, in the Context of Strengthening the Global Response to Threat of Climate Change, Sustainable Development, and Efforts to Eradicate Poverty*, Cambridge University Press, Cambridge, In Press, p. 9.
3 IPCC 2014a, 'Chapter 29: small islands' in IPCC 2014, *Climate Change 2014: Impacts, Adaptation and Vulnerability, Working Group II Contribution to the Fifth Assessment Report of the IPCC*, Cambridge University Press, Cambridge, p. 1617.
4 ICOMOS 2019, *The future of our pasts: engaging cultural heritage in climate action*, ICOMOS Climate Change and Cultural Heritage Working Group, Paris, France, p. 7.
5 Standing Committee on Environment, Communications, Information Technology and the Arts 2007, *Inquiry into Australia's Indigenous Visual Arts and Craft Sector*, Commonwealth of Australia, Canberra, p. 16
6 Standing Committee on Environment, Communications, Information Technology and the Arts 2007, p. 20.
7 Scott, M 2017, *Safekeeping: A Report on the Care and Management of Art Centre-Based Community Collections*, Arnhem, Northern and Kimberley Artists Aboriginal Corporation (ANKA), p. 4.
8 Scott 2017, p. 4.
9 It is worth noting that the publication on Gija views of climate change described in Chapter 3: Barney, G, Cann, C, Carrington, B et al 2013, *Jadagen Warnkan Barnden: Changing Climate in Gija Country*, Warmun Art Centre, Warmun, NT, Australia, was produced by the Gija-owned and managed Warmun Art Centre.
10 Samuel, R 1976, 'Local history and oral history', *History Workshop Journal*, Vol. 1, p. 99, as cited in Cosson, F 2017, 'The small politics of everyday life: local historical society archives and the production of public histories', *Archives & Records*, Vol. 38, No. 1, p. 50.
11 Questions from 'Climatic and Environmental Threats to Cultural Heritage: Issues for Communities', an electronic survey conducted by the authors during October and September 2021, targeting community heritage organisations such as local historical societies.
12 World Health Organisation (WHO) 2017, 'Bohol earthquake: one year on', *WHO News*, 2 July, accessed on 12 October 2021, https://www.who.int/philippines/news/feature-stories/detail/bohol-earthquake-one-year-on
13 Carlowicz, M & Schollaert Uz, S 2017, 'El Niño Pacific Wind and Current Changes Bring Warm, Wild Weather', *NASA Earth Observatory News*, 14 February, accessed on 2 April 2022, https://earthobservatory.nasa.gov/features/ElNino
14 Iese, V, Kiem, AS, Mariner, A, Malsale P, Tofaeono, T, Kirono, DGC, Round, V, Heady, C, Tigona, R, Veisa, F, Posanau, K, Aiono, F, Haruhiru, A, Daphne, A, Vainikolo, V & Iona, N 2021, 'Historical and future drought impacts in the Pacific islands and atolls', *Climatic Change*, Vol. 166, No. 19, p. 2.
15 lese et al 2021, p. 2.
16 Iese et al 2021, p. 4.
17 Iese et al 2021, p. 12.
18 Savage, A, Bambrick, H & Gallegos, D 2021, 'Climate extremes constrain agency and long-term health: a qualitative case study in a pacific small island developing state', *Weather and Climate Extremes*, Vol. 31, p. 6.
19 Savage, Bambrick & Gallegos 2021, p. 6.

20 Savage, Bambrick & Gallegos 2021, p. 9.
21 Iese et al 2021, p. 11.
22 Liu, C, Linde, A & Sacks, I 2009, 'Slow earthquakes triggered by typhoons', *Nature*, Vol. 459, pp. 833–6. https://doi.org/10.1038/nature08042
23 McGuire, B 2012, *Waking the Giant: How Changing Climate Triggers Earthquakes, Tsunamis and Volcanoes,* Oxford University Press, Oxford.
24 Cunneen, J 2022, '5 ways climate change increases the threat of tsunamis, from collapsing ice shelves to sea level rise', *The Conservation*, 24 January, accessed on 2 April 2022, https://theconversation.com/5-ways-climate-change-increases-the-threat-of-tsunamis-from-collapsing-ice-shelves-to-sea-level-rise-175247
25 Karm, M & Sarma, M 2021, 'Climate change, natural calamities and the triple burden of disease', *Nature Climate Change*, Vol. 11, pp. 796–7.
26 IPCC 2018, p. 5.
27 IPCC 2018, p. 10.
28 Zanotti, L, Ma, Z, Johnson, JL, Johnson, DR, Yu, DJ, Burnham, M & Carothers, C 2020, 'Sustainability, resilience, adaptation, and transformation: tensions and plural approaches', *Ecology and Society*, Vol. 25, No. 3.
29 Zanotti et al 2020.
30 Derissen et al 2011, Anderies et al 2013, Johnson et al 2018 in: Zanotti et al 2020.
31 Zanotti et al 2020.
32 Dr. A Labrador 2021, pers. comm., 23 October.
33 M Tissandier 2021, pers. comm., 27 October.
34 Zanotti et al 2020.
35 Zanotti et al 2020.
36 Thomsen, DC, Smith, TF, & Keys, N 2012, 'Adaptation or manipulation? Unpacking climate change response strategies', *Ecology and Society* Vol. 17. No. 3. See also: Smit et al 2000, Smith & Stirling 2010.
37 Zanotti et al 2020.
38 United Nations Office for Disaster Risk Reduction (UNDRR) 2015, *The Sendai Framework for Disaster Risk Reduction 2015–2030*, United Nations.
39 UNDRR 2015, p. 11.
40 UNDRR 2015, p. 25.
41 UNDRR 2015, p. 25.
42 United Nations Educational, Scientific and Cultural Organisation (UNESCO) 2017, *Towards Climate Change Resilience: Minimising Loss and Damage in Pacific SIDS Communities*, UNESCO, Fauntleroy, France, p. 5.

References

Carlowicz, M & Schollaert Uz, S 2017, 'El Niño Pacific wind and current changes bring warm, wild weather', *NASA Earth Observatory News*, 14 February, accessed on 2 April 2022, https://earthobservatory.nasa.gov/features/ElNino

Cosson, F 2017, 'The small politics of everyday life: local historical society archives and the production of public histories', *Archives & Records*, Vol. 38, No. 1, pp. 45–60, doi:10.1080/23257962.2016.1269646

Cunneen, J 2022, '5 ways climate change increases the threat of tsunamis, from collapsing ice shelves to sea level rise', *The Conservation*, 24 January, accessed on 2 April 2022, https://theconversation.com/5-ways-climate-change-increases-the-threat-of-tsunamis-from-collapsing-ice-shelves-to-sea-level-rise-175247

ICOMOS (International Council on Monuments and Sites) 2019, *The Future of Our Pasts: Engaging Cultural Heritage in Climate Action*, ICOMOS Climate Change and Cultural

Heritage Working Group, Paris, France, https://indd.adobe.com/view/a9a551e3-3b23-4127-99fd-a7a80d91a29e

Iese, V, Kiem, AS, Mariner, A, Malsale P, Tofaeono, T, Kirono, DGC, Round, V, Heady, C, Tigona, R, Veisa, F, Posanau, K, Aiono, F, Haruhiru, A, Daphne, A, Vainikolo, V & Iona, N 2021, 'Historical and future drought impacts in the Pacific islands and atolls', *Climatic Change*, Vol. 166, No. 19, pp. 1–24, doi:10.1007/s10584-021-03112-1

IPCC (Intergovernmental Panel on Climate Change) 2014a, 'Chapter 29: small islands' in IPCC 2014, *Climate Change 2014: Impacts, Adaptation and Vulnerability, Working Group II Contribution to the Fifth Assessment Report of the IPCC*, Cambridge University Press, Cambridge, pp. 1613–1654.

IPCC 2018, 'Summary for policymakers' in IPCC 2018, *Global Warming of 1.5°C. An IPCC Special Report on the Impact of Global Warming of 1.5°C above Pre-industrial Levels and Related Global Greenhouse Gas Emission Pathways, in the Context of Strengthening the Global Response to Threat of Climate Change, Sustainable Development, and Efforts to Eradicate Poverty*, Cambridge University Press, Cambridge, In Press, pp. 3–24.

Karm, M & Sarma, M 2021, 'Climate change, natural calamities and the triple burden of disease', *Nature Climate Change*, Vol. 11, pp. 796–797, doi:10.1038/s41558-021-01164-w

Liu, C, Linde, A & Sacks, I 2009, 'Slow earthquakes triggered by typhoons', *Nature*, Vol. 459, pp. 833–836, doi:10.1038/nature08042

McGuire, B 2012, *Waking the Giant: How Changing Climate Triggers Earthquakes, Tsunamis and Volcanoes*, Oxford University Press, Oxford.

Samuel, R 1976, 'Local history and oral history', *History Workshop Journal*, Vol. 1, pp. 191–208, https://www.jstor.org/stable/4288044

Savage, A, Bambrick, H & Gallegos, D 2021, 'Climate extremes constrain agency and long-term health: a qualitative case study in a Pacific Small Island Developing State', *Weather and Climate Extremes*, Vol. 31, pp. 1–11, doi:10.1016/j.wace.2020.100293

Scott, M 2017, *Safekeeping: A Report on the Care and Management of Art Centre-Based Community Collections*, Arnhem, Northern and Kimberley Artists Aboriginal Corporation (ANKA).

Smit, B, Burton, I, Klein, RJT & Wandel, J 2000, 'An anatomy of adaptation to climate change and variability', *Climatic Change*, Vol. 45, pp. 223–251, doi:10.1023/A:1005661622966

Smith, A & Stirling, A 2010, 'The politics of social-ecological resilience and sustainable socio-technical transitions', *Ecology and Society*, Vol. 15, No. 1, doi:http://www.ecologyandsociety.org/vol15/iss1/art11/

Standing Committee on Environment, Communications, Information Technology and the Arts 2007, *Inquiry into Australia's Indigenous Visual Arts and Craft Sector*, Commonwealth of Australia, Canberra.

Thomsen, DC, Smith, TF & Keys, N 2012, 'Adaptation or manipulation? Unpacking climate change response strategies', *Ecology and Society*, Vol. 17, No. 3, doi:10.5751/ES-04953-170320

UNDRR (United Nations Office for Disaster Risk Reduction) 2015, *Sendai Framework for Disaster Risk Reduction 2015–2030*, United Nations, https://www.preventionweb.net/files/43291_sendaiframeworkfordrren.pdf

UNESCO (United Nations Educational, Scientific and Cultural Organisation) 2017, *Towards Climate Change Resilience: Minimising Loss and Damage in Pacific SIDS Communities*, UNESCO, Fauntleroy, France, https://unesdoc.unesco.org/ark:/48223/pf0000260140.locale=en

World Health Organisation (WHO) 2017, 'Bohol earthquake: one year on', *WHO News*, 2 July, accessed on 12 October 2021, https://www.who.int/philippines/news/feature-stories/detail/bohol-earthquake-one-year-on

Zanotti L, Ma, Z, Johnson, JL, Johnson, DR, Yu, DJ, Burnham, M & Carothers, C 2020, 'Sustainability, resilience, adaptation, and transformation: tensions and plural approaches', *Ecology and Society*, Vol. 25, No. 3, pp. 1–4, doi:10.5751/ES-11642-250304

8

ISSUES FOR INSTITUTIONS: THE IMPERATIVE FOR HERITAGE ORGANISATIONS TO LEAD THE WAY

Robyn Sloggett and Marcelle Scott

There are many ways in which heritage institutions and peak body organisations engage with the climate crisis. These include managing risk and mitigating impact in areas for which they have responsibility; undertaking research for more effective decision-making; delivering education and advocacy initiatives; and by developing effective policies and strategies. They do this as peak bodies, public institutions, professional organisations and key stakeholders in the protection and preservation of heritage. This chapter examines the work of these institutions and organisations to identify trends in how organisations are thinking about the climate crisis, and to understand where and how gaps may exist in current work in this area. From the latter quarter of twentieth century, through to the first decade of the twenty-first-century climate change was predominately aligned to and subsumed under the topic of sustainability. This was the result of the interest in indoor environmental parameters that emerged from the late 1970s, in part focused by the increase in international blockbuster exhibitions towards the end of the twentieth century. This iteration of sustainability focussed on the management of environmental conditions for the protection of buildings and objects, and to address the escalating costs for air-conditioning. It was also prompted by the realisation in the Global North that institutions in Southeast Asia, Africa, Australia and the Pacific were dealing with a very different range of environments. Sustainability is still a key word in the responses of institutions and organisations to climate change, but the term is now more likely to include reference to, and describe actions that, relate specifically to mitigating increasing climate change.

This chapter also explores the articulated influence that has occurred as, at an individual level, heritage professionals (the archaeologists, curators, archivists, conservators and others who work in heritage organisations across the globe) are affected by climate change and its associated environmental impacts. As a result, this individual and 'bottom-up' interest has also translated into strategic frameworks, with

DOI: 10.4324/9781003163312-9

professional bodies playing a significant role in developing industry-wide approaches that also often strongly link across to approaches and responses from local communities. There is also 'top-down' influence, as non-government peak bodies such as UNESCO and the International Panel on Climate Change (IPCC) and national, state and local government organisations formulate policies and strategies. As a result, many institutions are aligning public programs, exhibitions, policy and programs with government climate change strategies and professional codes and statements.

The museum environment: changing roles and responsibilities

The museum environment as a place of security or risk has been a focus for collection managers and conservators since at least the mid-nineteenth century, when the National Gallery of London's *Report from the Select Committee on National Monuments and Works of Art, 16 June 1841* described the Gallery as 'wretchedly ventilated'.[1] The interest was focused in 1978 with the publication of Garry Thomson's *The Museum Environment*. Thomson urged caution in the implementation of overly prescriptive temperature and relative humidity parameters, noting in the final chapter of his book:

> There is something inelegant in the mass of energy-consuming machinery needed at present to maintain constant RH and illuminance, something inappropriate in an expense which is beyond most of the world's museums ... We cannot, of course, prophesy what will be developed, but I should guess that it will include means for stabilising the RH in showcases without machinery, use of solar energy for RH control in the tropics, improved building construction to reduce energy losses and extensive electronic monitoring.[2]

Despite the evidence, widespread implementation of Thomson's very practical prediction was slow, a reticence that is hard to fathom given his book would become a key source for managing indoor climates around the world.

In the last quarter of the twentieth century, interest in the public role of museums as social agents was growing, culminating in publications such as Stephen Weil's influential book *Making Museums Matter* (2002).[3] In 2007, Dr Richard J Hebda, acclaimed botanist, climate historian, curator of Botany and Earth History at the Royal British Columbia Museum (commonly referred to as the Royal BC Museum), wrote about climate change as an impetus for museums to refocus

> to become more socially relevant, not simply as a place to see and read about interesting things (and perhaps to strengthen our identity), but also to engage meaningfully in the most vital challenge of our time.[4]

Sarah Staniforth, then Heritage Properties Director of Museums and Collections at the UK National Trust, recast this as a focus on the sustainable museum. Bringing the internal discourse around environmental parameters with the more public-facing discourse around sustainability. Staniforth called for a 'Slow Conservation' movement, arguing that:

> Museums have been slow to prioritize reducing their environmental footprints. Indeed, many have considered that the significance of their collections justifies high energy consumption in achieving comfort conditions for their collections and visitors.[5]

By the second decade of the twenty-first-century concerns around what might constitute optimum temperature and relative humidity for collection institutions was overshadowed by the need of many national institutions and heritage organisations to respond as victims of rapid onset climate change-induced disasters. In January 2020 in Australia, the National Gallery of Australia was closed to protect the national collection from smoke damage from bushfires.[6] Three months earlier in October 2019 in the United States, the Getty Museum had closed as a result of wildfires.[7] Eighteen months later in August 2021, the iconic Archaeological Site of Olympia was threatened by fires a kilometre away.[8]

At the same time, cultural institutions in Germany and other parts of Europe were cleaning up after some of the worst flooding ever experienced. In the United States, in November 2021 *The New York Times*, under the banner headline, 'Saving History With Sandbags: Climate Change Threatens the Smithsonian' claimed 'Now, because of climate change, the Smithsonian stands out for another reason: Its cherished buildings are extremely vulnerable to flooding, and some could eventually be underwater'.[9] In the Pacific cultural institutions and heritage places are experiencing increasing challenges from cyclones and new challenges from rising sea levels. There is no longer a choice for many institutions to be involved in climate change. The climate crisis posits them as both agents and victims in a global catastrophe.

More recently, Marianne Tissandier, Head of Collections at the Museum of New Caledonia, expanded the focus to take account of how the impact of climate change on people's view of the environment is affecting their relationship to the museum, noting that:

> My main concern is watching how links with the objects are changing - the change in the way people see things because their environment is changing so the ways they are linked to the objects are changing … it's a concern for me that climate change is changing the way people have their relationship with their cultural heritage.[10]

In the face of what Sir David Attenborough described as 'the biggest threat to our security that human beings have ever faced',[11] this chapter provides a snapshot of how heritage institutions and organisations see their changing roles and responsibilities in response to the climate crisis.

Public positioning

In 2002 English Heritage commissioned the Centre for Sustainable Heritage at University College London to conduct a scoping study of climate change risks to the historic environment, and suitable mitigation and adaptation strategies. The project's final report provided a detailed analysis of the risks with evidence 'of a fragile heritage which could be damaged by much less than a catastrophic event'.[12] The report's author, Professor May Cassar, provided eight recommendations, outlining key steps and practices that address the concerns identified in the scoping study. The recommendation headings provide a series of sequential 'signposts' that could be designed and applied to a wide range of heritage types and locations. They are:

> Recommendation 1: Sector leadership on climate change
> Recommendation 2: Monitoring, management and maintenance
> Recommendation 3: Value and significance in managing climate change impacts
> Recommendation 4: Participation in the planning of other agencies
> Recommendation 5: Fully functional heritage information system
> Recommendation 6: Emergency preparedness
> Recommendation 7: Adaptation strategies and guidelines for historic buildings, archaeology, parks and gardens
> Recommendation 8: Buried archaeology and prediction maps (to inform management or conservation plans for archaeological landscapes or sites).[13]

Cassar also presented a paper, with co-author Robyn Pender, at the 2005 ICOM-CC Triennial Conference held in The Hague, highlighting the responsibility of those in the sector to share information, collaborate and influence policy. They identified the need to educate the public 'on the impact of climate change on cultural heritage and on the importance of cultural heritage as a climate change indicator', to record the impacts of climate change at a local level, and to have a mechanism in place to 'enable information, resources and good examples of adaptation' to be shared internationally.[14]

Two years later, the European Commission's *Noah's Ark: Global Climate Change Impact on Built Heritage and Cultural Landscapes* project (2004–2007) produced a vulnerability atlas of climate scenarios showing the threats to outdoor built heritage and landscapes, and adaptation strategies to assist 'stakeholders, policy makers and organisations to contemplate future pressures'.[15] The same year the Getty Conservation Institute convened the *Experts' Roundtable on Sustainable Climate Management Strategies* to draw together a range of perspectives on energy consumption and more environmentally and economically sustainable means of managing climates within cultural institutions, specifically in museums, galleries, libraries, and archives. Participants considered the

> current climate management strategies and emerging trends; the meaning of sustainability in relation to the preservation of cultural heritage; and whether cultural institutions such as museums, archives, and libraries *can or should* play a role in the debate about energy consumption [Emphasis added].[16]

Early in the proceedings, Sarah Staniforth reminded the group that the Trust was already thinking about climate change impacts on its properties, and had adopted the definition of conservation as 'the careful management of change' as they worked towards climate proofing the buildings. She identified two options:

> One is mitigation of climate change by reducing the amount of carbon dioxide added to the atmosphere through energy efficiency and considering the type of energy that we use. And the other aspect is adaptation. And adaptation has two aspects. In the day-to-day maintenance of the building, the lesson that we are already learning is that climate change will find out poor standards of preventive conservation. But the other aspect is the way in which we adapt buildings to cope with these extreme weather events.[17]

Conservation scientist and pioneer of low-energy museum and archive storage repositories, Tim Padfield, offered a vigorous criticism of the energy demands of modern museums:

> Air-conditioning is not the purpose of the museum. The purpose of the museum is to have curators, to have active people researching in archaeology and history … There are museums that are spending three-quarters of their income just on the air-conditioning, the engineers, the people who go around calibrating instruments, and so on … The idea that [museum buildings] should be good containers has simply vanished. We have a rather urgent business to bring some sanity into museum design.[18]

Soon after the International Institute for Conservation of Historic and Artistic Works (IIC) Congress, held in London, included roundtable discussions to consider the impact of climate change on museum and historic house collections. *Climate Change and Museum Collections* became the first of the *Dialogues for a New Century* series hosted by the IIC.[19]

By 2016, ICOMOS had established its Climate Change and Heritage Working Group (CCHWG), and published the report, 'Future of Our Pasts: Engaging Cultural Heritage in Climate Action' which aimed 'to develop a roadmap for heritage organizations to engage on climate change issues, to support the creation of new a doctrinal heritage text on climate change and cultural heritage, and to organise outreach to the scientific community on research gaps and opportunities'.[20] In 2020, the ICOMOS General Assembly declared a Climate and agreed to focus the Ecological Emergency, and associated Triennial Scientific Plan climate change. The role of the Climate Action Working group now concentrates on two objectives:

1. The establishment of an online resource portal or one-stop shop for the ICOMOS a wider heritage community, which will contain thematic resources including examples of best practice, relevant polices and guidance and case studies of stories from around the world.
2. To create an engagement plan to communicate, raise awareness and literacy and promote action amongst the ICOMOS community.

Other organisations such as the Southeast Asian Regional Centre for Archaeology and Fine Arts (SEAMEO-SPAFA), Pacific Island Museum Association (PIMA), World Archaeology Congress (WAC) and International Council on Archives (ICA) address climate change in conferences, programs and events. While a number have Standing Committees, Working Groups or Taskforces dedicated to the Disaster Risk Reduction and Disaster Risk Management, they do not appear to have a dedicated strategic focus on climate change.

In early March 2020, the Smithsonian American Art Museum (SAAM) and the Smithsonian's National Collections Program convened a meeting 'Stemming the Tide: Global Strategies for Sustaining Cultural Heritage Through Climate Change'. The meeting was framed by six categories of cultural heritage: Cultural Landscapes and Historic Urban Landscapes, Archaeological Sites, Built Heritage (Buildings and Structures), Cultural Communities, Intangible Cultural Heritage, and Museums and Collections. Dr Lonnie G. Bunch, Secretary of the Smithsonian, wrote that:

> It is time for the cultural sector to act deliberately and decisively. We must capitalize on our long legacy of environmental research and public engagement. We must marshal our resources and mobilize our vast networks to take on the challenge that will define the coming century.[21]

With this in mind, the sector will need to move beyond discussions about temperature and relative humidity parameters for indoor Heating, Ventilation and Air Conditioning systems (HVAC) towards active leadership roles and partnerships in local and global strategies. Sarah Sutton contends, however, although the level of activity in the sector demonstrates its capacity, a widespread commitment to action is lacking.[22] Sutton calls for

> professional associations to embrace and galvanize sector-wide action. As member-driven organizations, professional associations have traditionally taken their cues from inside the profession … this does not work with climate action. Associations must take their cues from the public, government, science, and lived experience. They must choose to lead from out front, not from behind. Our professional associations have the influence and

> credibility to mobilize action. They alone can elevate enough leaders, support enough of the followers, and encourage enough of the beleaguered in this work to lead widespread change.[23]

The need to understand the relationships between climate change and environmental parameters within heritage buildings and collecting institutions, the challenging cost to institutions in managing existing infrastructure or investing in additional infrastructure, and the need to provide leadership within an issue that is of global importance, continues to confront collecting institutions and peak bodies within the heritage sector. As the survey responses in the next section indicate for many organisations the focus on sustainability has not yet translated to discourse around climate change; although the alignment is often implied. It has also, particularly, not translated into the heritage sector playing a leadership role in discourse around the climate crises.

One reason why activity and capacity are evident, but coordinated leadership is less evident, might rest with what Dietrich Dörner described in *The Logic of Failure* as the inability to make appropriate decisions in complex systems. In referencing Dörner, the Canadian Conservation Institute (CCI) identified the problem as

> the general blindness problem … when people try to make decisions in complex systems that have either very slow feedback (in our case, very slow deterioration) or very infrequent feedback (in our case, rare disasters).[24]

Surveying the sector

In 2021, in order to understand in some detail, the responses to and ideas of climate change from heritage institutions, organisations and peak bodies, the authors undertook a set of qualitative surveys across cultural heritage organisations and peak bodies in Australia and the Pacific, which explored:

- Whether the institution, organisation or body had a public position on the impacts of climate change.
- Whether there was strong support amongst staff, membership, and board for a statement regarding the impacts of climate change, any issues involved in producing a statement, and any positive or negative outcomes from publishing such a statement.
- A description of the process taken to develop a statement on climate change.
- Any obligation to align statements with those of other organisations and/or government.
- Whether it was clear that members of the organisation have adopted the intent of the statement, and whether there were any activities within the organisation, or any other initiatives that the respondents have been involved in that relate to climate change initiatives.

The results indicated a general interest in climate change but reported a variety of responses. Some organisations and institutions were heavily committed to action, including through committee activity, knowledge sharing, adaptation, and formal commitments. Others, while understanding climate change as a critical issue, remained unsure how to best approach it. Further in-depth interviews were held with Marianne Tissandier from the Museum of New Caledonia and Dr Ana Labrador, then Deputy Director of the National Museum of the Philippines and member of the International Council of Museum's Standing Committee for Museum Definition, Prospects and Potentials.

Three areas emerged as key to understanding the role of institutions, organisations and peak bodies in the climate crisis. These were whether the organisation had a public position on climate change; what actions and activities they focused on in any work related to climate change; and what role policy and strategic planning played in responses to climate change.

Public positions

While some organisations had clear public positions on climate change, many did not. In some cases, organisations identified a broader and more generic position. Arnhem, Northern and Kimberley Artists Aboriginal Corporation (ANKA), which is a peak body for Aboriginal artists and Aboriginal-owned and managed art centre, highlighted the indivisible relationship between climate change, art, culture and place identifying '… homelands as the heartlands of Indigenous art and culture and therefore they are of profound significance to these areas'.[25] Australian Institute for the Conservation of Cultural Material (AICCM), which is the peak body for conservators in Australia, ratified its *Environmental Guidelines for Australian Cultural Heritage Collections* in 2018[26] and works through its Sustainable Collections Committee and Disaster Preparedness Committee to lead project work in these two related areas, but has no public statements on climate change.[27] Similarly, the Art Gallery of New South Wales (AGNSW) has a sustainability platform developed through the sustainability committee and which is 'been actively adopted by staff'.[28]

On the other hand, the National Trust Australia (Victoria) (NTAV) argued that the climate crisis 'requires urgent and coordinated action' which requires NTAV to '… use our platform to exchange knowledge and information with our members, stakeholders, and partners to create more resilient communities and achieve a safe climate future.[29] The Museum of Applied Arts and Science (MAAS) in Sydney published a public position on climate change and reported strong support from staff and the board 'for the development of a statement regarding the impacts of climate change'.[30] Queensland Art Gallery and Gallery of Modern Art (QAGOMA) has adopted a highly proactive position on climate change with a Sustainability Policy and related procedures, with sustainability in strategic and operational plans.[31] At the Museum of New Caledonia key individuals within the organisation, including the Museum's director, and the human resources director who has a direct interest in sustainability, ecological issues, and climate change, have led an institutional focus on climate change. The New Caledonian government's green objectives, and the new museum development, have provided the museum with the opportunity to take a leadership role in this area.[32] In the United States, the Foundation for Advancement in Conservation reported on the Foundation's active working group on climate change,[33] and the Center for Architectural Conservation/ Weitzman School of Design/University of Pennsylvania reported on its working group that focused on vulnerability frameworks.[34]

Actions and activities

While many organisations did not identify a public statement on climate change, nevertheless responses indicated most have sustainable practices aimed at reducing greenhouse gas emissions and waste. NTAV identified: 'Building preservation, adaptive reuse and conservation methodologies … [as] forms of climate action', noting the 'environmental benefits of retaining and adaptively reusing existing building stock'.[35] The respondent for Museums and Galleries New South Wales (MGNSW) noted that many small galleries were incorporating building strategies and developing resources 'to minimise environmental impact'.[36] Federation of Australian Historical Societies (FAHS), the peak body for about 1000 community history and heritage organisations, identified digitisation as an areas of response to climate change into which FAHS has 'put considerable effort into encouraging and facilitating the digitisation of collections, partly through membership of GLAM Peak'.[37] The Australasian Society for Historical Archaeology (ASHA) noted that its members were very aware of the impacts of climate change and natural disasters as a result of 'the Christchurch earthquake, the 2019/2020 bushfire in NSW, followed by floods in NSW and Queensland, as well as the (man-made) destruction of Juukan Gorge', and were active educators and lobbyists in this area.[38] A respondent in Singapore noted that within her institution promoting practical activity including 'recycling and creative use of packaging waste' had been very effective.

The National Museum of the Philippines utilises its 15 museums across the Philippines for workshops with local scientists, and shared spaces that minimise the building footprint of the museum as some responses to climate change. Currently, a site in Iloilo is being co-shared with the National Historical Commission, which is considered a useful model for future initiatives. In order to mitigate the high cost of managing environmental controls across the National Museum's three buildings, the Museum has increased the use of double glazing, and to cope with increased level of rain the roof size has been doubled.[39] The Museum of New Caledonia's current redevelopment and expansion has increased the recognition of the need to incorporate sustainable materials into the building design. For example, the museum's new storage areas have solar rooftop panels, resulting in a 20-times reduction in the electricity costs.[40]

Tasmanian Museum and Art Gallery (TMAG) accounts for climate change in its disaster planning, within exhibitions and public talks. Although the only respondent to report on this, the role TMAG scientists in dealing with the effects of climate change regularly in their research, is highly relevant to the role played by such organisations in regard to climate change.[41] The National Gallery of Victoria's (NGV) Conservation Department hosts regular Collection Environment meetings that are attended by staff from Conservation and Engineering and Facilities, and is responsible for regular review of policies and documents such as loan documentation. There are also 'frequent information and training sessions related to recycling initiatives'. The Conservation Department has staged testing program and has undertaken research over several years to assist cultural change around perceived risks to collections relating to the hesitancy, noting concern in some aspects of the organisation for a loosening of environmental guidelines.[42] AGNSW has adopted new approaches to recycling, and like the Museum of New Caledonia is embedding sustainability principles in the design of the new building. The AGNSW also reported on a program of active knowledge sharing with other organisation to develop new learning and enable improvements. Overall 'staff are enthusiastic and committed to making changes at all levels'.[43]

QAGOMA has set its own target of 2024–2025 to be Climate Active – carbon neutral, and offers an active program of exhibitions, film programs and public programs that focus on climate change and sustainability. It has developed initiatives to support sustainable practices to help address climate change, including waste separation into recycling and composting streams which in 2020 'diverted 90 tonnes (of what would have been landfill) to managed waste streams'; sourcing of specialist timbers from sustainable and renewable supplies, with used construction materials passed to Men's Shed and other charity organisations; and the use of 100% organic waste packaging and composting of all food waste. QAGOMA has also adopted the AICCM 'Environmental Guidelines' of which Amanda Pagliarino, who is head of Conservation, was an author and key driver, and the BIZOT Green Protocol.[44]

One respondent from Singapore made the point that the COVID-19 pandemic diverted staff attention and institution resources 'to respond and manage the impact of pandemic', highligting the issue of cascading impacts.[45]

Policy and strategic frameworks

The role of articulated policies and strategies from local level, such as local committee and departments organisations, through to government policy, strategy and legislation was an important factor in whether organisations were actively and publicly engaged in responses to climate change, or not. Conversely, the impetus from staff and members of organisations for the organisations to be actively responding to climate change, was also a factor in decision organisations were making.

The peak body for the archives sector, the Australian Society of Archivists (ASA), reported strong support amongst its members for the development of statements, procedures and activities that respond to climate change. A strategic aim of the ASA is sustainability, and this includes assessing the environmental impact of the organisation and its members.[46] FAHS also indicated strong support for statements, procedures and activities that addressed climate change amongst its members, and provides advice to members on climate change. While the work of Blue Shield does not directly factor climate change into its emergency response framework, both ASA and FAHS identified Blue Shield as a key organisation with which they engaged.

Michael Varcoe-Cocks, Associate Director of Conservation at the National Gallery of Victoria (NGV), noted the importance of the organisation having a public position on sustainability that links to the Victorian State Government's Climate Strategy. According to Varcoe-Cocks

> this has changed the way staff work and think. Complacency is no longer seen as acceptable and resistance to change (by staff) in a long-established institution can be better supported by the official stance.[47]

As a member of the Bizot Group, the NGV supported the Bizot Green Protocol and the Council of Australian Art Museum Directors' endorsement of the AICCM Environmental Guidelines.

QAGOMA's Sustainability Committee aligns to the Queensland State Government position. QAGOMA's respondent, Amanda Pagliarino, pointed to the particularly active role adopted by individuals, including the QAGOMA Director, Chris Saines, who was Chair of Council of Australian Art Museum Directors (CAAMD) which endorsed the adoption of the AICCM Environmental Guidelines and the BIZOT Green Protocol for Loans.[48] QAGOMA has supported Pagliarino in her leadership activities in researching, publishing and disseminating information on sustainability and climate change. This, in turn, has enabled the AICCM to play a leadership role in developing more climate friendly environmental parameters in Australian collecting institutions. In 2018, Pagliarino was awarded a Churchill Fellowship 'To develop sustainable collection environment guidelines for Australian cultural organisations'.[49]

In the Pacific, the Museum of Samoa and the Solomon Islands' National Museum both identified working with UNESCO as impactful. For the Solomon Islands, this was through the recent consultation workshop convened to discuss the protection of underwater cultural heritage, an area at high risk from severe weather events, as part of the ratification of the 2001 UNESCO Convention on the Protection of the Underwater Cultural Heritage.[50] The Museum of Samoa noted the application of resilient skills and ideas for addressing issues raised by climate change that were highlighted in the workshop, 'Revitalization of Indigenous Architecture and Sustainable Building Skills in the Pacific', held in Samoa in 2014 training program led by International Training Center for Intangible Cultural Heritage in the Asia-Pacific Region (CRIHAP) under the auspices of UNESCO.

Key messages

There were a number of additional 'take home' messages provided by respondents. The respondent from the Museum of Samoa expressed interest in supporting, and being involved in, common dialogue about heritage and climate change 'as it is global issue', which they identified as relevant in the preservation work of the Museum because of how 'vulnerable island states [are] to climate change'.[51] The respondent suggested value in more training opportunities and project funds that are specifically focused on preservation and conservation in the Pacific Islands that are vulnerable to climate change, and particularly for preservation work for tangible and intangible heritage.[52]

The respondent from the Solomon Islands National Museum noted that impacts of climate change on cultural, historical and archaeological heritages are not factored into the National Climate Change program[53] because climate change in the Solomon Islands is mostly seen as an environmental issue.

Ana Labrador from the National Museum of the Philippines pointed to the day-to-day effects of the climate crisis, noting that

> the climate crisis here is really terrible ... you know it's been raining almost nonstop every day for over a month and then we also have issues ... in terms of building maintenance ... the higher humidity is causing issues on the building materials so even if we have a building code I think we have to adjust that now because of stronger typhoons and an extreme heat.[54]

She expressed the need to make sustainability 'an everyday thing for our staff to just simply switch off lights when there's no one in the room and things like that', noting that policies do not necessarily empower staff to take action.[55]

The respondent from the NTAV cited the US National Trust for Historic Preservation's 2011 document 'The Greenest Building: Quantifying the Value of Building Reuse' as useful in highlighting the important role of 'building reuse and retrofits [that] substantially reduce climate change impacts'.[56] They cited the length of time of 10–80 years for a new building, built at standards which are '30% more efficient than an average performing existing building' to overcome the negative impacts that result from the construction process.

Marianne Tissandier, also raised the issue of preserving intangible heritage in New Caledonia, commenting that 'the seasons are getting more intense and shifting a little in the calendar'.[57] Tissandier spoke to the role of museums as both scientific and cultural organisations, with an important role to play in understanding social change resulting from the climate crisis. She explained, 'we have a lot of objects that are related to the cycle of the young, and the correct traditional calendar is based on the cycle of the young', which is now the focus for discussion as people recognise changes in the climate. As a result, people are identifying changes in the intangible parts of the culture, and the way they understand objects and the value of the objects is also changing.[58]

Activating cultural heritage organisations in the fight against the climate crisis

What is clear, from both the literature on heritage and climate change and from the responses to the surveys is that for many in the heritage sector two areas, adaptation and sustainability, appear as the most comfortable places from which to mount attacks against the climate crisis. These, in turn, align to the strategy of adaptation and adaptative reuse in the built heritage sector, and preventive conservation in the collections sector. In 2016, the Canadian Conservation Institute, in collaboration with ICCROM (International Centre for the Study of the Preservation and Restoration of Cultural Property), produced a manual 'The ABC Method: a risk management approach to the preservation of cultural heritage', which sought to provide a resource across various types of heritage to help 'assess the risks and deterioration processes affecting our heritage asset, and then to act to reduce them as effectively as possible, given the available resources'.[59] The process for doing so is structured around two ongoing activities: communicate and consult, and monitor and review; and five sequential steps: '1. Establish the context; 2. Identify risks; 3. Analyze risks; 4. Evaluate risks; 5. Treat risks'.[60]

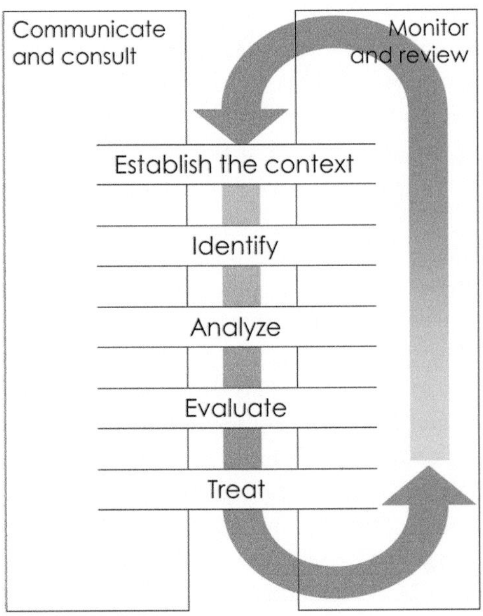

FIGURE 8.1 The risk management cycle. ©Government of Canada, Canadian Conservation Institute.

Fig. 8.1 as principles, these two activities and five steps provide a useful framework from which to develop a broader response to climate crises. In particular, the specificities of the Ten Agents of Deterioration, located under Step 2 'Identify Risks', provide areas that can be mapped and measured for impact and change. These Ten Agents are:

1. Physical forces;
2. Thieves, vandalism and other criminal activity;
3. Fire, and other catastrophic events;
4. Water, Flood;
5. Pests;
6. Pollutants/contaminants;
7. Light (esp UV);
8. Temperature changes;
9. RH changes;
10. Custodial neglect and dissociation.[61]

Many of these agents will be more aggressively damaging as the climate crisis increases in severity and embed in the future. The case could also be made for two additional agents to be added to this list in order to properly map measure their impacts. They are:

11. Development;
12. Climate change.

Naming the climate crisis as a serious threat to the future of tangible and intangible heritage provides a broader context for the specific risks associated with this crisis. It is a necessary first step in ensuring that the first-order problem, the climate crisis, rather than the second-order response, sustainability, sits at the core of planning, management and presentations for every conference, congress, symposium and meeting for the next decade.

As survey respondents and the literature indicate, implementing change at the local level within organisations, institutions and peak bodies generally receives positive responses amongst staff and members. Local responses within organisations are also embraced by managers and boards when they align to government or other national or international climate action or sustainability programs or policies. As the data in this chapter has also shown, there are very few examples of organisations and peak bodies working across sectors to share ideas and information.

One way of doing this is to concentrate on developing generic processes that can be shared across various types of heritage organisations. Life Cycle Assessments (LCA) are one example of a methodology that is useful to understand risk and to monitor effective responses.[62] Utilising an LCA methodology in conjunction with other risk management strategies, such as the ABC Method, ensures that the impacts of choices that might deal with immediate risks do not have unintended longer-term consequences. One particularly pertinent example has been provided by Pendergrass et al who examined the negative consequences of digital preservation programs, and reported on the massive environmental footprint from digitisation in the archives sector. Citing the Jevons Paradox, 'an ecological economics concept that explains how greater efficiency of a process leads to increased use of the resources in that process', they noted that the 'paperless office'

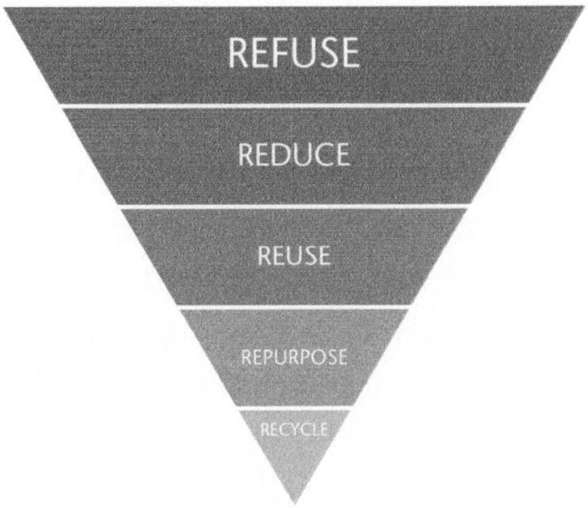

FIGURE 8.2 The five Rs principle of waste management

has, paradoxically, led to an increased use in paper.[63] They explain that by understanding the impacts of the Information, Communication and Technology (ICT) cycle a more comprehensive assessment of environmental impact of digitisation can be made. Embedding LCA as part of daily work routines helps support clearly understood, properly managed and effective decision-making.

There are other principles that are also useful. Embedding the five Rs principles of waste management: refuse, reduce, reuse, repurpose and recycle into every aspect of professional decision-making provides an effective 'lived leadership' model for heritage organisations (Fig. 8.2).

Being able to understand and assess the environmental footprint for standard activities (such as collection management; site reporting; documentation; storage; conservation and preservation; digitisation; loans; exhibitions; and so on), and being able to access data that supports impactful local decision making would be a significant initiative which could be led by a consortium of peak bodies. This requires baseline data to be identified, collected, ingested and shared. Project Drawdown provides information to support effective decision-making to help mitigate climate change across a range of activities. The Drawdown methodology enables various sectors to measure alternatives to help make more informed and effective decisions.[64] Scorecard Assessments can also be a valuable and practical tool in providing advice to individuals and organisations across the heritage sector.[65] Like the Drawdown model, they provide baseline data to support decisions about energy use. Both Drawdown and Scorecard Assessment processes have the potential to be modified for heritage-specific purposes. For example, they can help understand options that might best mitigate cultural memory loss resulting from climate crisis-induced migration, or what materials are most suitable for adaptive reuse. In order to do so, extensive research programs need to be activated.

There is also opportunity to bring together institutional capacity and local knowledge to provide better emergency preparedness and emergency responses. Ana Labrador cited lessons learned from locals on some of the smaller islands in the Bohol region where, after Typhoon Haiyan, almost no lives were lost because local knowledge enabled appropriate disaster responses to occur.[66]

Activating organisations in the fight against the climate crisis

This chapter has identified three key areas where heritage institutions and organisations are actively engaged in climate change. These are research, community engagement and advocacy. The survey responses and initiatives discussed in this chapter demonstrate examples of research and methodologies that illustrate how heritage organisations can respond to the challenges posed by the climate crisis.[67] Institutions are actively engaged in public advocacy, publishing explicit statements about their position on sustainability and climate change, developing and delivering public programs related to the climate crisis, and establishing professional dialogues. Despite these leadership activities much more needs to be done if heritage organisations, institutions and peak bodies are to properly realise their potential as leaders in the fight to ensure that heritage is protected from the risks that result from the climate crisis.

Notes

1 *Report from the Select Committee on National Monuments and Works of Art, 16 June 1841,* paragraph 2658 cited in Saunders, D 2000, 'Pollution and the National Gallery', *National Gallery Technical Bulletin*, Vol. 21, p. 77.
2 Thomson, G 1978, *The Museum Environment*, Butterworths, London, p. 267.
3 Weil, S 2002, *Making Museums Matter*, Smithsonian Institution, Washington DC.
4 Hebda, RJ 2007, 'Museums, climate change and sustainability', *Museum Management and Curatorship*, Vol. 22, No. 4, p. 329.
5 Staniforth, S 2010, 'Slow conservation', *Studies in Conservation,* Vol. 55, No.2, p. 76.
6 Cascone, S 2020, 'Smoke from Australia's devastating wildfires has forced the closure of the National Gallery in Canberra', *artnet*, 6 January, accessed on 5 April 2022, https://news.artnet.com/art-world/wildfire-smoke-closes-national-gallery-australia-1746144
7 Cascone, S 2019, 'The Getty Museum closes its doors as wildfires rage in California, but says the art is safe inside its fireproof building', *artnet*, 29 October, accessed 2 November 2021, https://news.artnet.com/art-world/getty-museum-closed-wildfires-1691749
8 Mills, E 2021, 'Greek wildfires threaten UNESCO World Heritage site of Olympia', *Museums Association Journal*, 10 August, accessed on 2 January 2022, https://www.museumsassociation.org/museums-journal/news/2021/08/greek-wildfires-threaten-unesco-world-heritage-site-of-olympia/#
9 Flavelle, C 2021, 'Saving history with sandbags: climate change threatens the Smithsonian', *New York Times*, 25 November, accessed on 5 April 2022, https://www.nytimes.com/2021/11/25/climate/smithsonian-museum-flooding.html
10 Tissandier, M 2021, *Climatic and Environmental Threats to Cultural Heritage: Issues for Institutions* [interview], Project Reference No: 2021-20948-19279-5, University of Melbourne.
11 United Nations 2021, *Sir David Attenborough Speech to the UN Security Council* [video], YouTube, 23 February, accessed on 5 April 2022, 03:33, https://youtu.be/MaweqwsN62k
12 Cassar, M 2005, *Climate Change and the Historic Environment*, Centre for Sustainable Heritage, UCL, London, UK, p. 1.
13 Cassar 2005, pp. 2–3.
14 Cassar, M & Pender, R 2005, 'The impact of climate change on cultural heritage: evidence and response' in *ICOM-CC 14th Triennial Meeting Preprints, The Hague, 12–16 September 2005*, Vol. II, p. 614.

15 European Commission 2007, *Noah's Ark: Global Climate Change Impact on Built Heritage and Cultural Landscapes*, Report #SSPI-CT-2003–501837, Community Research and Development Information Service (CORDIS), p. 2, accessed on 12 February 2021, https://cordis.europa.eu/project/id/501837/reporting
16 Getty Conservation Institute (GCI) 2007, *Experts Roundtable on Sustainable Climate Management Strategies,* edited transcript, Tenerife, Spain, pp. 1–2, accessed on 12 February 2021, https://www.getty.edu/conservation/our_projects/science/climate/roundtable_transcript.pdf
17 GCI 2007, p. 5.
18 GCI 2007, pp. 7–8.
19 International Institute for Conservation of Historic and Artistic Works (IIC) 2008, *Climate Change and Museum Collections*, edited transcript, 17 September 2008, accessed on 10 September 2021, https://www.iiconservation.org/sites/default/files/dialogues/climate-change-en.pdf
20 Climate Change and Heritage Working Group, 2019 'The future of our pasts: engaging cultural heritage in climate action. Outline of climate change and cultural heritage', ICOMOS, p. 6, accessed on 12 September 2021, https://openarchive.icomos.org/id/eprint/2459/1/CCHWG_final_print.pdf
21 Bunch, LG 2021, 'Foreword' in R Rushfield (ed) *Stemming the Tide: Global Strategies for Sustaining Cultural Heritage through Climate Change*, Smithsonian Scholarly Press, Washington DC, p. v.
22 Sutton, S 2020, 'The evolving responsibility of museum work in the time of climate change', *Museum Management and Curatorship*, Vol. 35, No. 6, pp. 618–35.
23 Sutton 2020, p. 633.
24 Dörner, D 1996, *The Logic of Failure: Recognizing and Avoiding Error in Complex Systems*, translated by R Kimber & R Kimber, Metropolitan Books, New York, as cited in CCI/IIC & ICCROM, 2016, *The ABC Method: A Risk Management Approach to the Preservation of Cultural Heritage,* Canadian Conservation Institute, Ottawa, CA, p. 33.
25 ANKA (Arnhem, Northern and Kimberley Artists) 2021, *Climatic and Environmental Threats to Cultural Heritage: Issues for Institutions* [survey], Project Reference No: 2021-20948-19279-5, University of Melbourne.
26 AICCM 2019, 'Environmental guidelines', accessed on 15 July 2020, https://aiccm.org.au/conservation/environmental-guidelines/
27 AICCM 2021, *Climatic and Environmental Threats to Cultural Heritage: Issues for Institutions* [survey], Project Reference No: 2021-20948-19279-5, University of Melbourne.
28 AGNSW 2021, *Climatic and Environmental Threats to Cultural Heritage: Issues for Institutions* [survey], Project Reference No: 2021-20948-19279-5, University of Melbourne
29 NTAV (National Trust of Australia, Victoria) 2021, *Climatic and Environmental Threats to Cultural Heritage: Issues for Institutions* [survey], Project Reference No: 2021-20948-19279-5, University of Melbourne.
30 Randall, B 2021, *Climatic and Environmental Threats to Cultural Heritage: Issues for Institutions* [survey], Project Reference No: 2021-20948-19279-5, University of Melbourne.
31 Pagliarino, A 2021, *Climatic and Environmental Threats to Cultural Heritage: Issues for Institutions* [survey], Project Reference No: 2021-20948-19279-5, University of Melbourne.
32 Tissandier, M 2021, *Climatic and Environmental Threats to Cultural Heritage: Issues for Institutions* [interview], Project Reference No: 2021-20948-19279-5, University of Melbourne.
33 Foundation for Advancement in Conservation 2021, *Climatic and Environmental Threats to Cultural Heritage: Issues for Institutions* [survey], Project Reference No: 2021-20948-19279-5, University of Melbourne.
34 Center for Architectural Conservation 2021, *Climatic and Environmental Threats to Cultural Heritage: Issues for Institutions* [survey], Project Reference No: 2021-20948-19279-5, University of Melbourne.
35 NTAV 2021.

36 Museums and Galleries NSW 2021, *Climatic and Environmental Threats to Cultural Heritage: Issues for Institutions* [survey], Project Reference No: 2021-20948-19279-5, University of Melbourne; see, for example, https://mgnsw.org.au/sector/resources/online-resources/sustainability/ and https://magsq.com.au/facility-management-development/

37 FAHS (Foundation of Australian Historical Societies) 2021, *Climatic and Environmental Threats to Cultural Heritage: Issues for Institutions* [survey], Project Reference No: 2021-20948-19279-5, University of Melbourne;

GLAM Peak is a coalition of peak bodies in the galleries, libraries, archives and museum (GLAM) sector and includes key research institutions, which, in October 2021, endorsed the United Nations' Sustainable Development Goals, see https://glampeak.org.au/

38 ASHA (Australasian Society for Historical Archaeology) 2021, *Climatic and Environmental Threats to Cultural Heritage: Issues for Institutions* [survey], Project Reference No: 2021-20948-19279-5, University of Melbourne.

39 Labrador 2021.

40 Tissandier 2021.

41 O'Connell, J 2021, *Climatic and Environmental Threats to Cultural Heritage: Issues for Institutions* [survey], Project Reference No: 2021-20948-19279-5, University of Melbourne.

42 Varcoe-Cocks, M 2021, *Climatic and Environmental Threats to Cultural Heritage: Issues for Institutions* [survey], Project Reference No: 2021–20948-19279-5,2021-20948-19279-5, University of MelbourneMelbourne.

43 AGNSW 2021, *Climatic and Environmental Threats to Cultural Heritage: Issues for Institutions* [survey], Project Reference No: 2021-20948-19279-5, University of Melbourne.

44 AICCM 2019; BIZOT Green Protocol 2014, accessed on 20 June 2021, https://cimam.org/sustainability-and-ecology-museum-practice/bizot-green-protocol/

45 Anonymous Respondent 2021, *Climatic and Environmental Threats to Cultural Heritage: Issues for Institutions* [survey], Project Reference No: 2021-20948-19279-5, University of Melbourne.

46 ASA (Australian Society of Archivists) 2021, *Climatic and Environmental Threats to Cultural Heritage: Issues for Institutions* [survey], Project Reference No: 2021-20948-19279-5, University of Melbourne.

47 Varcoe-Cocks 2021.

48 Pagliarino 2021.

49 Churchill Trust 2018, 'Amanda Pagliarino, QLD 2018', accessed on 11 September 2021, https://www.churchilltrust.com.au/fellow/amanda-pagliarino-qld-2018/

50 Solomon Islands National Museum 2021, *Climatic and Environmental Threats to Cultural Heritage: Issues for Institutions* [survey], Project Reference No: 2021–20948-19279-5,2021-20948-19279-5, University of Melbourne.

51 Museum of Samoa 2021, *Climatic and Environmental Threats to Cultural Heritage: Issues for Institutions* [survey], Project Reference No: 2021-20948-19279-5, University of Melbourne.

52 Museum of Samoa 2021.

53 Solomon Islands National Museum 2021.

54 Labrador 2021.

55 Labrador 2021.

56 NTAV 2021.

57 Tissandier 2021.

58 Tissandier 2021.

59 CCI (Canadian Conservation Institute), in collaboration with ICCROM (International Centre for the Study of the Preservation and Restoration of Cultural Property), 2016, *The ABC Method: a risk management approach to the Preservation of Cultural Heritage*, Canadian Conservation Institute, Ottawa, CA. p. 8.

60 CCI/IIC & ICCROM 2016, p. 20. The manual notes on p. 19 that: 'This manual is structured around a risk management cycle that was found originally in the Australian / New Zealand Standard for Risk Management (AS/NZS 4360:2004) and that is now part of the ISO 31000:2009, Standard for Risk Management'.
61 CCI/IIC & ICCROM 2016, p. 70.
62 Udo, H & van Rooijen, M 2005, *Life Cycle Approaches: The Road from Analysis to Practice*, UNEP/SETAC Life Cycle Initiative, Paris, accessed on 10 January 2022, https://www.lifecycleinitiative.org/wp-content/uploads/2012/12/2005%20-%20LCA.pdf
63 Pendergrass, K, Walker, S, Walsh, T & Alagna, L 2019, 'Toward environmentally sustainable digital preservation', *The American Archivist*, Vol. 82, No. 1, p. 172.
64 Project Drawdown 2022, 'Table of solutions', accessed on 5 February 2022, https://www.drawdown.org/solutions/table-of-solutions
65 Victorian State Government 2022, 'What is a Scorecard assessment?', accessed on 10 March 2022, https://www.victorianenergysaver.vic.gov.au/save-energy-and-money/get-a-home-energy-assessment/getting-an-assessment
66 Labrador 2021.
67 See Pagliarino, A & Meredith, A 2021, 'Climate change and risks to Australian cultural collections: maps and workflow resource, *AICCM*, accessed on 2 April 2022, http://aiccm.org.au/wiki/climate-adaptation; Sutton 2020; Pendergrass et al 2019.

References

AGNSW 2021, *Climatic and Environmental Threats to Cultural Heritage: Issues for Institutions* [survey], Project Reference No: 2021-20948-19279-5, University of Melbourne.

AICCM (Australian Institute for the Conservation of Cultural Materials) 2019, 'Environmental guidelines', accessed 15 July 2020, https://aiccm.org.au/conservation/environmental-guidelines/

AICCM (Australian Institute for the Conservation of Cultural Materials) 2021, *Climatic and Environmental Threats to Cultural Heritage: Issues for Institutions* [survey], Project Reference No: 2021-20948-19279-5, University of Melbourne.

ANKA (Arnhem, Northern and Kimberley Artists) 2021, *Climatic and Environmental Threats to Cultural Heritage: Issues for Institutions* [survey], Project Reference No: 2021-20948-19279-5, University of Melbourne.

ASHA (Australasian Society for Historical Archaeology) 2021, *Climatic and Environmental Threats to Cultural Heritage: Issues for Institutions* [survey], Project Reference No: 2021-20948-19279-5, University of Melbourne.

ASA (Australian Society of Archivists) 2021, *Climatic and Environmental Threats to Cultural Heritage: Issues for Institutions* [survey], Project Reference No: 2021-20948-19279-5, University of Melbourne.

Bunch, LG 2021, 'Foreword' in R Rushfield (ed) *Stemming the Tide: Global Strategies for Sustaining Cultural Heritage through Climate Change*, Smithsonian Scholarly Press, Washington, DC, pp. v–vi.

Cascone, S 2019, 'The Getty Museum closes its doors as wildfires rage in California, but says the art is safe inside its fireproof building', *artnet*, 29 October, accessed 2 November 2021, https://news.artnet.com/art-world/getty-museum-closed-wildfires-1691749.

Cascone, S 2020, 'Smoke from Australia's devastating wildfires has forced the closure of the National Gallery in Canberra', *artnet*, 6 January, accessed on 5 April 2022, https://news.artnet.com/art-world/wildfire-smoke-closes-national-gallery-australia-1746144.

Cassar, M 2005, *Climate Change and the Historic Environment*, Centre for Sustainable Heritage, UCL, London, England.

Cassar, M & Pender, R 2005, 'The impact of climate change on cultural heritage: evidence and response' in *ICOM-CC 14th Triennial Meeting Preprints, The Hague, 12–16 September 2005*, Vol. II, pp. 610–616.

CCI/IIC & ICCROM 2016, *The ABC Method: A Risk Management Approach to the Preservation of Cultural Heritage*, Canadian Conservation Institute, Ottawa, CA.

Center for Architectural Conservation 2021, *Climatic and Environmental Threats to Cultural Heritage: Issues for Institutions* [survey], Project Reference No: 2021-20948-19279-5, University of Melbourne.

Churchill Trust 2018, 'Amanda Pagliarino, QLD 2018', https://www.churchilltrust.com.au/fellow/amanda-pagliarino-qld-2018/

Dörner, D 1996, *The Logic of Failure: Recognizing and Avoiding Error in Complex Systems*, translated by R Kimber & R Kimber, Metropolitan Books, New York.

European Commission 2007, *Noah's Ark: Global Climate Change Impact on Built Heritage and Cultural Landscapes*, Report #SSPI-CT-2003-501837, Community Research and Development Information Service (CORDIS), https://cordis.europa.eu/project/id/501837/reporting

Flavelle, C 2021, 'Saving history with sandbags: climate change threatens the Smithsonian', *New York Times*, 25 November, accessed 5 April 2022, https://www.nytimes.com/2021/11/25/climate/smithsonian-museum-flooding.html

FAHS (Foundation of Australian Historical Societies) 2021, *Climatic and Environmental Threats to Cultural Heritage: Issues for Institutions* [survey], Project Reference No: 2021-20948-19279-5, University of Melbourne.

Getty Conservation Institute (GCI) 2007, *Experts Roundtable on Sustainable Climate Management Strategies*, edited transcript, Tenerife, Spain, https://www.getty.edu/conservation/our_projects/science/climate/roundtable_transcript.pdf

Hebda, RJ 2007, 'Museums, climate change and sustainability', *Museum Management and Curatorship*, Vol. 22, No. 4, pp. 329–336, doi:10.1080/09647770701757682

International Institute for Conservation of Historic and Artistic Works (IIC) 2008, *Climate Change and Museum Collections*, edited transcript, 17 September 2008, accessed on 10 September 2021, https://www.iiconservation.org/sites/default/files/dialogues/climate-change-en.pdf

Labrador, A 2021, *Climatic and Environmental Threats to Cultural Heritage: Issues for Institutions* [interview}, Project Reference No: 2021-20948-19279-5, University of Melbourne.

Mills, E 2021, 'Greek wildfires threaten UNESCO world heritage site of Olympia', *Museums Association Journal*, 10 August, accessed on 2 January 2022, https://www.museumsassociation.org/museums-journal/news/2021/08/greek-wildfires-threaten-unesco-world-heritage-site-of-olympia/#.

Museum of Samoa 2021, *Climatic and Environmental Threats to Cultural Heritage: Issues for Institutions* [survey], Project Reference No: 2021-20948-19279-5, University of Melbourne.

Museums and Galleries NSW 2021, *Climatic and Environmental Threats to Cultural Heritage: Issues for Institutions* [survey], Project Reference No: 2021-20948-19279-5, University of Melbourne.

NTAV (National Trust of Australia, Victoria) 2021, *Climatic and Environmental Threats to Cultural Heritage: Issues for Institutions* [survey], Project Reference No: 2021-20948-19279-5, University of Melbourne.

O'Connell, J 2021, *Climatic and Environmental Threats to Cultural Heritage: Issues for Institutions* [survey], Project Reference No: 2021-20948-19279-5, University of Melbourne.

Pagliarino, A 2021, *Climatic and Environmental Threats to Cultural Heritage: Issues for Institutions* [survey], Project Reference No: 2021-20948-19279-5, University of Melbourne.

Pagliarino, A & Meredith, A 2021, 'Climate change and risks for Australian cultural collections: maps and workflows resource, *AICCM*, accessed on 2 April 2022, https://aiccm.org.au/wiki/climate-adaptation/

Pendergrass, K, Walker, S, Walsh, T & Alagna, L 2019, 'Toward environmentally sustainable digital preservation, *The American Archivist*, Vol. 82, No. 1, pp. 165–206, doi:10.17723/0360-9081-82.1.165

Project Drawdown 2022, 'Table of solutions', accessed on 5 February 2022, https://www.drawdown.org/solutions/table-of-solutions

Randall, B 2021, *Climatic and Environmental Threats to Cultural Heritage: Issues for Institutions* [survey], Project Reference No: 2021-20948-19279-5, University of Melbourne.

Saunders, D 2000, 'Pollution and the National Gallery , *National Gallery Technical Bulletin*, Vol. 21, pp. 77–94, doi:http://www.jstor.org/stable/42616143

Solomon Islands National Museum 2021, *Climatic and Environmental Threats to Cultural Heritage: Issues for Institutions* [survey], Project Reference No: 2021-20948-19279-5, University of Melbourne.

Staniforth, S 2010, 'Slow conservation', *Studies in Conservation*, Vol. 55, No. 2, pp. 74–80, doi:10.1179/sic.2010.55.2.74

Sutton, S 2020, 'The evolving responsibility of museum work in the time of climate change', *Museum Management and Curatorship*, Vol. 35, No. 6, pp. 618–635, doi:10.1080/09647775.2020.1837000.

Thomson, G 1978, *The Museum Environment*, Butterworths, London.

Tissandier, M 2021, *Climatic and Environmental Threats to Cultural Heritage: Issues for Institutions* [interview], Project Reference No: 2021-20948-19279-5, University of Melbourne.

Udo, H & van Rooijen, M 2005, *Life Cycle Approaches: The Road from Analysis to Practice*, UNEP/SETAC Life Cycle Initiative, Paris, accessed on 10 January 2022, https://www.lifecycleinitiative.org/wp-content/uploads/2012/12/2005%20-%20LCA.pdf

UNESCO 2021, 'UNESCO World Heritage Centre statement regarding the fires in Greece', *UNESCO World Heritage Convention News*, 9 August, accessed on 2 November 2021, https://whc.unesco.org/en/news/2329

United Nations 2021, *Sir David Attenborough Speech to the UN Security Council* [video], YouTube, 23 February, accessed on 5 April 2022, 03:33, https://youtu.be/MaweqwsN62k

Varcoe-Cocks, M 2021, *Climatic and Environmental Threats to Cultural Heritage: Issues for Institutions* [survey], Project Reference No: 2021-20948-19279-5, University of Melbourne.

Victorian State Government 2022, 'What is a Scorecard assessment?', https://www.victorianenergysaver.vic.gov.au/save-energy-and-money/get-a-home-energy-assessment/getting-an-assessment

9

CONSERVATION EDUCATION AND CLIMATE CHANGE: WHAT WOULD A RESILIENCE-BASED CONSERVATION CURRICULUM LOOK LIKE?

Marcelle Scott and Robyn Sloggett

The climate crisis is now no longer manageable over one generation. It is also no longer possible to keep global temperature rises under 1.5°C, and many of the impacts of this current generation's inability to address climate change are now a tragic legacy for next generations. It is sobering to remember that for a number of species, there will be no next generation. The burden for future generations is therefore heavy, both emotionally and practically. This chapter examines the responsibility of current educators to incorporate active climate change responses into cultural heritage teaching programs. It also explores some initiatives that are being incorporated into current teaching programs. While the focus is on the discipline of cultural materials conservation, which teaches predominantly within Bachelor undergraduate degrees and Master by coursework programs, the findings and principles incorporated into this chapter are relevant to all heritage professionals and heritage experts who are engaged in educating their next generations.

In 1922, educationalist John Dewey wrote:

> In order that education of the young be efficacious … it is not necessary for adults to have a formulated definite ideal of some better state … What is necessary is that habits be formed which are more intelligent, more sensitively percipient, more informed with foresight, more aware of what they are about, more direct and sincere, more flexibly responsive than those current. Then they will meet their own problems and propose their own improvements.[1]

While tertiary education in the twenty-first century has a diverse age demographic, the principle of providing an education that results in intelligent, percipient, informed, aware, direct, sincere responsive students is a critical proposition for educators working to address the impacts of the climate crisis. Nowhere was this perhaps

DOI: 10.4324/9781003163312-10

as apparent as with the Student Climate Action Network when students turned their back on their classrooms to draw attention to, and educate the older generations about, the impacts of the climate crisis. In this way, students demonstrated both the extent of their anxiety and concern for their own future and that of the planet, but also the resilience they found in working together to respond to this threat.

Dewey's exhortation for 'efficacious' education is very similar to calls to provide content that will help to develop resilience thinking in students. Such content can, as Josef Ploner argues, provide 'a different and more focussed lens through which to understand students' values, expectations, capabilities, anxieties and hopes in the course of their studies (and beyond)'.[2] Ploner's work can also inform the structure of a resilience-based curriculum that, rather than being 'derive(d) from prescribed, top-down teaching practices' is based on 'participatory and action-oriented educational practices that acknowledge students' own capacity building'.[3] Fumiyo Kagawa's study of education delivered in response to emergency situations provides an insight into some of the specific aspects of this approach. She emphasises the critical importance of local expertise, adding that 'what is needed is a sea change in the direction of flow of knowledge both at theoretical and at practical levels'.[4] Kagawa cites Naomi Klein, who argued that those engaged in education in emergency situations should be very mindful that 'people in emergencies are their own experts in their experience' and aware of their right to be part of a communal recovery.[5] Highlighting the risks of a top-down approach, Kagawa adds that 'even those who go into disaster situations with the best intentions … run the risk of seeking short-term fixes at the expense of doing longer term permanent damage'.[6] Grounded studies such as these demonstrate the central role of community in disaster response and resilience, reinforcing their place in curriculum development and delivery.

Student-informed approaches

Like all heritage preservation studies, cultural materials conservation knowledge, discourse and praxis are fundamentally interdisciplinary. Like most disciplines, its knowledge framework has undergone a rapid expansion in recent decades to extend content areas to include global concerns of sustainability, climate change and human rights. The combination of multiple streams of knowledge from the humanities and the sciences is the basis of conservation research and subject-based pedagogies. In this mix of hard science and abstract philosophical inquiry, cultural and social knowledge and perspectives are central, embodying the 'human purpose' of why society requires and enables conservators to do the work they do. Given that conservation requires this social pact then conservation education is a two-way learning process, with conservators, like other heritage experts, needing to learn from the individuals and communities who hold the cultural knowledge, while also educating the public about conservation.

Every two years since 2012, the University of Melbourne has sought feedback from staff and students on their views of the University's sustainability initiatives, via a voluntary and anonymous online survey. In 2020, the survey received over 1200 responses, which demonstrated a high level of awareness of the issues and their belief that the university has a responsibility to include climate change impacts and sustainability measures in all aspects of its work. The following selected responses from students capture these convictions.

In response to the question 'Do you have any suggestions for how the University of Melbourne can better communicate/engage with the University community on sustainability?' one student responded:

> Support programs that integrate accessible options for anyone to get involved, actively contribute and learn about sustainability from more than just the white perspective ....[7]

Two responses to the question 'Do you have any ideas about how to improve the environmental practices of the University of Melbourne?' highlight students' awareness of the complexity and interdisciplinary nature of the subject area. In one student's opinion, 'The University should keep sustainability at the forefront of their curriculum, research, values, culture, community'. Another suggested the University could 'Improve student understanding through application of cross disciplinary sustainability information integration for all students, and provide incentives and resources for students to partake'.[8]

Another respondent expressed a degree of dissatisfaction with current practices, calling on the university to:

> Insist on stronger standards of action on campus, including purchasing and supply. For example, insist that food cannot be sold in a single use container. The University should purchase products made from recycled soft plastics. If those products don't match the University's aesthetic then they should coinvest in design projects through engineering/industrial design to rectify that.[9]

The students' perspectives reflect David W Orr's rationale for the need to rethink educational approaches when he urges educators to recognise that they are working with student cohorts who are aware that they 'must begin the great work of repairing as much as possible, the damage done to the earth in the past two hundred years of industrialization'.[10] Without evidence of this awareness being embedded in the curriculum, there is the danger that a lack of relevancy and credibility will begin to erode effective teaching and learning.

Rigorous curriculum design is the basis of sound pedagogy. Good curriculum design demonstrates a responsiveness to social change.[11] There is also substantial evidence that student's learning is enhanced when they have been actively engaged in the subject content of the program.[12] In a crowded curriculum, adding responses to urgently needed social change becomes a complex undertaking.

Developing a climate crisis responsive curriculum

In examining climate change education and communication, Edgar Gonzalez-Gaudiano and Pablo Meiera-Cartea acknowledge the complex task of educating multiple publics and learning communities on climate change. They ask 'Do we need to know the carbon cycle and the laws of thermodynamics in order to accept the need to change the type of light bulb used?'.[13] Their point is not that the science is unimportant, but that an 'exclusively scientific' curriculum can miss opportunities to engage with the social context and experience of climate change in 'peoples' everyday lives'.[14] Instead, they argue 'We need a pedagogical strategy that is based on scientific knowledge, and social experience' [emphasis added].[15] Such interdisciplinary discourse and content is fundamental to conservation precepts, and is reflected in the pedagogical principles of most cultural materials conservation teaching programs, delivered through the topics of ethics, risk management, materials science, advocacy and community engagement.

One of the issues in developing a climate crisis responsive curriculum is, as the preceding chapter indicates, that there are not yet agreed approaches amongst peak bodies and institutions. Some have clear and embedded policies and procedures that address climate change, others have a more eclectic approach taking part in forums and short-term programs, and others are embedded in organisations that are still working through how to incorporate climate crisis responses into their systems of management and practice. While such a developing and fluid environment presents many opportunities to identify new initiatives and cutting-edge practices, it also makes it difficult to assess best practice. In turn, this makes the choice about what to include in the curriculum less straightforward. Another complexity, identified in the literature review undertaken by Fatorić and Seekamp, is that research into heritage and climate change remains piecemeal and uncoordinated.[16] This also has implications for research-led pedagogy, as there is much baseline research that remains to be done, and will be done in a highly politicised, dynamic and fluid environment. At the same time, this situation provides exceptional opportunities for early career researchers to take on effective, pioneering research on climate change.

As is the case with so many aspects of the topics discussed in this book, context is all important, and this remains the case for curriculum design and content. Carolin Kreber summarises the importance of contextualising content as a way of enhancing student learning and identifies success as

> defined by students not only learning lots about a subject, but knowing how they learned it and why what they learned matters to their understanding of and interaction with the world around them. As students acquire certain bodies of knowledge, procedures, conceptual tools, practices and even values within the particular context of our various subjects, their learning is context-specific, but also context-transcendent.[17]

Another aspect of curriculum design is course evaluation, both by examining the students' performance and that of their teachers and the teaching program. The type of assessment and the quality and timing of feedback given to students are some of the most critical parts of a curriculum, and of the overall student experience. Research undertaken by the authors, and based on a 2005 study of the first teaching cycle of the newly established Masters by Coursework in Cultural Materials Conservation at the Grimwade Centre for Cultural Materials Conservation, the University of Melbourne, showed that students can become overwhelmed by the volume of information they are required to synthesise. This was particularly the case in the early part of the program, and was exacerbated by the use of unfamiliar language and technical terms.[18] Not surprisingly, given the general anxiety and excitement students feel at the start of a new course, this research also found that if students are provided with more explicit information about the intended learning outcomes, and if the links between learning outcomes, assessment tasks and grading criteria are explicitly identified and discussed, then the students' performance and confidence increased.[19] As Jane Henderson and Phil Parkes, who teach the Bachelor of Science and Master of Science degrees in Conservation and Collection Care at Cardiff University, show 'assessment shapes students' learning', and that 'If conservation training is to be relevant to the communities that we serve, then their priorities for learning should correlate with the methods of assessment'.[20]

In this respect, having definitions that are clearly agreed across disciplinary domains is essential. As Chapter 8 demonstrates, galleries, libraries, archives and museums (the GLAM sector) and the conservation sector that supports them have focused for some time on the museum environment when discussing 'sustainability'. This has occured, despite the fact that the risks and responses to climatic and environmental change-induced disasters have been widely discussed across the GLAM sector and the conservation profession. The broad umbrella of 'sustainability' has received limited critical discussion of what the term really means for conservation discourse, planning, and, for that matter, pedagogy. In order to graduate students with the knowledge, skills, cognitive awareness, confidence and professional flexibility to be able to think critically and to apply their expertise within nuanced and rapidly changing contexts, then pedagogies and curricula need to be able to rely on a shared understanding of terminology related to climate change.

In 2006, Professor May Cassar, Director of University College London's Institute for Sustainable Heritage, opened a European Commission meeting of conservation educators in Vantaa, Finland. Examining the value of preventive conservation Cassar argued that sustainability aligns and supports conservation's ethical and humanistic practice, making the point that:

> The central aim of sustainability is to achieve an acceptable quality of life for the world's population, combined with economic growth of communities, without depleting or damaging the natural resources needed to sustain future generations. The same may be said of preventive conservation and its relationship with cultural heritage![21]

Three years later, she wrote 'One of the strongest sustainability arguments for heritage conservation is that material conservation is an inherently waste-avoidance activity'. She added that 'This argument is entirely consistent with a common definition of sustainability as the reduction of environmental impact by not consuming non-renewable resources'.[22] The articulation of these views by a leader in the field laid the groundwork for a radical expansion of conservation's precepts and practices to include sustainability as a key part of new curriculum development. The challenge remains, however, to find the time and space within an already crowded conservation curriculum to incorporate these values and practices in a meaningful way.

Graduate attributes

Graduate Attribute statements enable students to assess whether a particular program is likely to be a good fit with their own learning styles and career aspirations. Graduate attributes also enable education programs to publicly identify their values, intentions and aspirations. An examination of published Graduate Attributes for a range of cultural materials conservation programs in Europe, North America and Australia was undertaken to ascertain whether it was possible to determine the extent to which climate responses were incorporated into these aspirational and positioning statements. The publication of Graduate Attributes is not universal so a range of other statements including Learning Outcomes and Values Statements were also examined.[23] Only one conservation program specifically mentioned sustainability in its statement. This was the UCLA/Getty Conservation Program with a Mission Statement and Values which included Collaboration; Diversity, Equity and Inclusion; and Sustainability. A number of programs identified citizenship as defining characteristics of their graduate attributes. In Australia, both the University of Canberra and the University of Melbourne identified citizenship; Canberra categorised attributes under three headings: professional, global citizen, and lifelong learners, while the University of Melbourne's Grimwade Centre listed academic distinction, active citizenship, and integrity and self-awareness as defining features.

Several programs identified both soft and hard skills. Queens University in Canada included Creativity and innovation; Perseverance; Independence and experience as a collaborative worker; Awareness and understanding of sound ethical practices, social responsibility, responsible research, and cultural sensitivity; and Leadership; as well as more practical outcomes such as: Project management. The Values Statement from the University of Delaware identified 'empathy, cultural identity, inclusion, and social connection', and practical and interdisciplinary outcomes that aimed to educate 'professionals skilled in the analysis, treatment, and preservation of cultural artifacts by connecting disciplines in the arts, humanities, and sciences'. The University of Applied Arts Vienna (Universität für angewandte Kunst Wien) listed knowledge, international profession, communication and collaboration, communication and presentation, and networks. A similar set of outcomes was identified by the University of Barcelona and included skills in projects; physical assessment; multidisciplinary teamwork; research; and publishing. The Courtauld Institute noted a set of key attributes that included the ability to communicate effectively orally and in writing; intellectual independence and maturity; self-discipline and self-direction; respect for the views of others; project management; and ability to work in a team, collaborate and share resources.

This cursory review of graduate attributes statements across conservation education programs helps to identify how current aspirations to build the next generation of conservators may allow for the development of conservation advocates whose professional voice will make a difference and enhance the ability of heritage experts to fight climate change effectively. These public statements also define the position of education programs in relation to the climate crisis.

Conservation educators

While statements of graduate attributes and learning outcomes are one way of examining commitments to responding to climate change, a more nuanced understanding can be gained from examining what kind of climate change relevant content might be either being developed, or being embedded, in the curriculum. To find out more conservation educators were invited to share their views about the inclusion of sustainability and climate change topics in their current and/or future conservation programs via an online survey which included the following questions:

1. What level of qualification does your course offer?
2. In your view, are the impacts of climate change relevant to conservation curricula?
3. Please describe why you answered this way.
4. In your teaching program are topics related to climate change and/or sustainability included?
5. Please describe what topics relating to climate and sustainability are taught. How are these topics delivered? Are the topics part of compulsory subjects, or electives or both?
6. What are the impediments to including these topics in your conservation program?
7. What types of topics are incorporated into research programs? How are these topics incorporated into research programs?
8. What are the impediments to including sustainability and climate change as topics in your research program?[24]

The response was prompt and enthusiastic, with general agreement about the relevance and vocational purpose of including climate change in their program, which in one respondent's case is 'to inform our decision-making and to be embedded in our practice'.[25] In addition, a number of respondents noted the enthusiasm of their students and their future potential to lead the sector in sustainable practices, and to be 'actors for change in relation to climate change and environmental justice'.[26]

These and the following responses from conservation educators show a general shift in emphasis in the curriculum from purely theoretical to practical, achievable, practical and affordable responses that acknowledge both local and global perspectives.

At the UCLA/Getty Interdepartmental Program in the Conservation of Cultural Heritage which delivers both Masters and PhD programs, staff members recognise the numerous ways in which the conservation field has contributed to the climate crisis, such as 'through the use and disposal of toxic chemicals and non-sustainable materials and technologies' and by arguing 'for unsustainable museum practices including the maintenance of stringent temperature and RH levels'. In a response to the challenges of climate change, the program is increasing teaching in topics relating to 'environmental, social, economic, and cultural sustainability', and is introducing greener technologies into lab courses.[27]

Academic staff at Uppsala University Sweden identified the need for a climate crisis responsive curriculum, singling out a course in their Master's coursework degree titled 'Climate for Culture' that blends critical perspectives on heritage with the applied and technical research that is being conducted at the University on energy efficiency and indoor climate in historic buildings.[28]

The University of Glasgow's MPhil in Textile Conservation has a strong emphasis on sustainable conservation practices, as well as social, economic and cultural sustainability. The course has had practical outcomes for its graduates, one of whom worked with staff to examine how best 'to embed sustainability across the programme and to make this sustainable in the curricula and beyond the programme'.[29]

The National University of Singapore was one of only two respondents to link conservation education to the United Nations' Sustainable Development Goals, asserting that 'Conservation is [an] indispensable part of UN's 17 SDGs especially Goal 11; Global Warming is real, carbon neutrality is imperative, demolition is not the way to go'.[30] The Master's program embeds research related to climate change in 'two core compulsory modules (out of 3), 6 out of 10 stream modules, and some elective modules. Some of the topics are Adaptive Reuse, Disaster Risk Mitigation and Management, Conservation Policy and Methodology for Sustainable Development, etc'. They also identified topics in the Master's program that were 'related to UN's 17 SDGs, Adaptive Reuse, Zero Energy or Super Low Energy strategies, Vernacular wisdom, etc' and noting that the alignment is based on the fact that: 'Conservation is the management of change and shall be focused on human/community well-being that fulfills five-in-one principles: environmental sustainability, cultural authenticity, social inclusivity, economic viability, and physical integrity'.[31]

A respondent from the Winterthur/University of Delaware Program in Art Conservation (WUDPAC) Master's Degree (Research) argued that conservation programs need to consider climate change because a reliance on chemicals and expendable supplies means: 'Our carbon foot-print is large … It is important for us to think about how we reduce our reliance on toxic chemicals and minimize waste'. They added that as well as identifying a specific role with communities 'We also need to better understand how we can assist our coastal institutions when natural disasters strike'.[32] The respondent also identified disaster preparedness as an area of interest for students, including risks from 'increased storms due to climate change and the effects of sea level rise'.

Patricia Smithen from the Master of Art Conservation Program, Queen's University, Canada, has clear views on the topic, commenting that:

> In our field, we are seeing the impact of climate change with respect to disaster response to environmental incidents, increased awareness of sustainability practices in our own laboratory activities and also in the cultural heritage sector. It is a top global threat and should be addressed in all curricula.[33]

She explained that 'Queen's provides a mandatory course on Conservation Principles which deals with ethics, preventive conservation, and sustainability' and that 'Laboratory classes include discussions of sustainability and long-term impact of practices and materials, and teaches practices regarding safe use and disposal of chemicals'. Topics for lectures have also included disaster and emergency response. Smithen identified two recent student research projects as progressing relevant climate change-related research including one on 'conditions which limit or cause mold growth on modern paints' research'. Smithen also noted increased job creation in preventive conservation roles and suggested that 'a survey of preventive conservators might illuminate the shifts in policy, response and job growth which could be attributed to climate change'.

Nicole Tse, Discipline Chair for the University of Melbourne's Grimwade Centre felt, as a conservator and conservation educator, that the Centre had a professional responsibility to engage students 'in big picture discussions, using our knowledge to support discussions about identity and our skills to provide access to material culture are critical', with topics that aim to build knowledge that is socially and culturally representative.[34]

Each subject has a sustainability statement that outlines goals and acts to 'ground sustainable habits'. Tse identified Sustainable Development Goal 11.4 as aligning to the overall aims of the program, and then specific goals for individual subjects such as SDG 6 (Clean water and sanitation), SDG 7 (Affordable and clean energy), SDG 10 (Reduced inequalities), SDG 12 (Responsible consumption and production) and SDG 17 (Partnerships), and noted that students are beginning to report to these goals.

The responses from conservation educators, students and others included in this chapter suggest a growing readiness and capacity within the education sector for pro-active responses to climate change threats. They indicate an interest in building stronger foundations for effective climate crisis responses within the curriculum and allow us to envision what current and possible future resilience-based curricula might look like.

Investing in climate change pedagogy

Education about climate change takes place in an active and contested space. Student activity and advocacy was clearly demonstrated as an important driver in this respect. This in turn opens up the possibility for a more coordinated, student-led approach across education programs within the heritage education sector. Both current activity, and suggestions for future activity, included commentary on exploring and sharing information and resources, and identified current progress in addressing sustainability within specific programs. As the previous chapter indicated, professional bodies are actively progressing agendas that seek to contribute to, and in cases address, the challenges posed by the climate crisis. What appears as a gap is a more structured approach that can support professional bodies, institutions and higher education programs through research-led, evidence-based best practice.

Graduate attributes present the aspirations of tertiary education and research programs for how their students will be enabled to operate as professionals and as future researchers. A cursory examination of graduate attributes or associated statements across a number of conservation education programs in North America, Europe and Australia identifies how current aspirations to build the next generation of conservators may allow for the development of conservation advocates whose professional voice will make a difference to this critical issue. For this reason, an assessment of how these attributes support the development of students and graduates as effective advocates for the need to act responsibly, responsively and quickly to the climate crisis is warranted.

The majority of tertiary education programs include research capability as a core skill and research projects as a core outcome of their programs.

Approaches to building resources, for the heritage sector and for community heritage experts who are at the front line of the war on climate change, that respond to needs, are coordinated to fill gaps, and are accessible in ways that are needed, require an assessment of current research and research outcomes. This is a major, but important piece of work, given the increasingly shorter time frame that human beings have to confront the climate crisis, and the failure of responses to date. There is no longer the luxury of choosing what might be of interest, the focus needs to be on what is required, and required immediately.

Diversity in curriculum development and content is discussed within the sector, but there is considerably more work to be done examing programs for benchmarking two-way learning that is interdisciplinary and cross-cultural. Understanding what makes successful cross-cultural and interdisciplinary education is a program of pedagogical research that has the ability to substantially shift the ability of the heritage sector to respond to the challenges of climate change.

As this, and the previous chapters have shown, in recognising that the climate crisis requires responses that reach across cultural and disciplinary divides, education programs need to invest in understanding and supporting the value of Elders' knowledge as a significant part of education resourcing.

Notes

1 Dewey, J 1922, *Human Nature and Conduct*, Henry Holt & Co, New York, reprint, Dover Publications, Mineola, New York, 2002, p. 128.
2 Ploner, J 2011, *Promoting Students' 'Resilient Thinking' in Diverse Higher Education Environments*, Technical Report, The Higher Education Academy – Centre for Sociology, Anthropology, Politics (C-SAP), UK, p. 8.
3 Ploner 2011, p. 2.
4 Kagawa, F 2009, 'Learning in emergencies: defense of humanity for a livable world' in F Kagawa & D Selby (eds) *Education and Climate Change: Living and Learning in Interesting Times*, Routledge, London, p. 118.
5 Klein, N 2007, *The Shock Doctrine: The Rise of Disaster Capitalism*, Metropolitan Books, New York, as cited in Kagawa 2009, p. 117.
6 Miller, T 2005, 'Foreword' in Plan International 2005, *Children and the Tsunami: Engaging with Children in Disaster Response, Recovery and Risk Reduction*, Plan Ltd, Thailand, p. 2 as cited in Kagawa 2009, p. 118.
7 University of Melbourne (UoM) 2020, 'Sustainability Survey 2020 Final Report', https://sustainablecampus.unimelb.edu.au/__data/assets/pdf_file/0003/3590031/2020-Sustainability-Survey-Report.pdf, p. 59.
8 UoM 2020, p. 65.
9 UoM 2020, p. 67.
10 Orr, DW 1994, *Earth in Mind: On Education, Environment and the Human Prospect*, Island Press, Washington, DC, p. 26.
11 James, R & Baldwin, G 2002, *Nine Principles Guiding Teaching and Learning*, Centre for the Study of Higher Education, The University of Melbourne, p. 16.
12 Scott, M 2016, *Tradition and Innovation: Building the Conservation Discipline in Australia,* PhD Thesis, The University of Melbourne, Australia, p. 196. See also Coates 2008.
13 González-Gaudiano, E & Meira-Cartea, P 2010, 'Climate change education and communication: a critical perspective on obstacles and resistances' in F Kagawa & D Selby (eds) *Education and Climate Change: Living and Learning in Interesting Times*, Routledge, London, p. 31.
14 González-Gaudiano & Meira-Cartea 2010, p. 31.
15 González-Gaudiano & Meira-Cartea 2010, p. 31.
16 Fatorić, S & Seekamp, E 2017, 'Are cultural heritage and resources threatened by climate change? A systematic literature review', *Climatic Change*, Vol. 142, No. 1–2, pp. 227–54.
17 Kreber, C 2009, 'Supporting student learning in the context of diversity, complexity and uncertainty' in C Kreber (ed) *The University and Its Disciplines*, Routledge, New York and London, p. 4.
18 Scott, M 2008, 'Conservation interdisciplinarity and pedagogical implications' in J Bridgland (ed) *Preprints of the ICOM-CC 15th Triennial Conference New Delhi India 22–26 September 2008*, ICOM-CC, Paris, p. 122.
19 Scott 2008, p. 122.
20 Henderson, J & Parkes, P 2014, 'Do methods of assessment accurately reflect the priorities of conservation training?' in J Bridgland (ed) *ICOM-CC 17th Triennial Conference Preprints*, Melbourne, Australia, 17–19 September 2014, ICOM, Paris, p. 1.

21 Cassar, M 2006, 'Value of preventive conservation', presentation to a meeting at EVTEK Institute of Art & Design, Vantaa, Finland, accessed on 20 October 2021, https://bit.ly/3Ch1InJ.
22 Cassar, M 2009, 'Sustainable heritage: challenges and strategies for the twenty-first century', *APT Bulletin: Journal of Preservation Technology*, Vol. 40, No. 1, p. 6.
23 While not all institutions post details on graduate attributes online statements were available from University of Canberra, Graduate Certificate in Heritage Materials; University of Melbourne, Masters by Coursework in Cultural Materials Conservation; University of Applied Arts Vienna (Universität für angewandte Kunst Wien), MA Cultural Heritage Conservation and Management; University of Barcelona, Master's degree in the Direction of Conservation and Restoration Projects; Queen's University, Kingston, Master of Art Conservation; Cardiff University, MSc Conservation Practice; The Courtauld Institute of Art, MA Conservation of Easel Paintings; University of Delaware, Winterthur/University of Delaware Program in Art Conservation; UCLA/Getty Conservation Institute, Master's Program in the Conservation of Cultural Heritage; University College London, Principles of Conservation MA.
24 Questions from 'Climatic and Environmental Threats to Cultural Heritage: Education', an electronic survey conducted by the authors during 2021.
25 Thompson, K 2021, *Climatic and Environmental Threats to Cultural Heritage: Education* [survey], Project Reference No: 2021-20655-18206-3, University of Melbourne.
26 Tyler, L 2021, Climatic and Environmental Threats to Cultural Heritage: Education [survey], Project Reference No: 2021-20655-18206-3, University of Melbourne.
27 Wharton, G 2021, Climatic and Environmental Threats to Cultural Heritage: Education [survey], Project Reference No: 2021-20655-18206-3, University of Melbourne.
28 Leijonhufvud, G 2021, Climatic and Environmental Threats to Cultural Heritage: Education [survey], Project Reference No: 2021-20655-18206-3, University of Melbourne.
29 Thompson 2021.
30 Widodo, J 2021, Climatic and Environmental Threats to Cultural Heritage: Education [survey], Project Reference No: 2021-20655-18206-3, University of Melbourne.
31 Widodo, J 2021.
32 Department of Art Conservation, University of Delaware 2021, Climatic and Environmental Threats to Cultural Heritage: Education [survey], Project Reference No: 2021-20655-18206-3, University of Melbourne.
33 Smithen, P 2021, Climatic and Environmental Threats to Cultural Heritage: Education [survey], Project Reference No: 2021-20655-18206-3, University of Melbourne.
34 Tse, N 2021, Climatic and Environmental Threats to Cultural Heritage: Education [survey], Project Reference No: 2021-20655-18206-3, University of Melbourne.

References

Cassar, M 2006, 'Value of preventive conservation', presentation to a meeting at EVTEK Institute of Art & Design, Vantaa, Finland, accessed on 20 October 2021, https://bit.ly/3Ch1InJ

Cassar, M 2009, 'Sustainable heritage: challenges and strategies for the twenty-first century', *APT Bulletin: Journal of Preservation Technology*, Vol. 40, No. 1, pp. 3–11, https://www.jstor.org/stable/27650524

Coates, H 2008, *Attracting, Engaging and Retaining: New Conversations about Learning*, Australasian Student Engagement Report: Australasian Survey of Student Engagement, Australian Council for Educational Research Ltd, Camberwell, Victoria.

Department of Art Conservation, University of Delaware 2021, *Climatic and Environmental Threats to Cultural Heritage: Education* [survey], Project Reference No: 2021-20655-18206-3, University of Melbourne.

Dewey, J 1922, *Human Nature and Conduct*, Henry Holt & Co, New York, reprint, Dover Publications, Mineola, New York, 2002.

Fatorić, S & Seekamp, E 2017, 'Are cultural heritage and resources threatened by climate change? A systematic literature review', *Climatic Change*, Vol. 142, No. 1–2, pp. 227–254, doi:10.1007/s10584-017-1929-9

González-Gaudiano, E & Meira-Cartea, P 2010, 'Climate change education and communication: a critical perspective on obstacles and resistances' in F Kagawa & D Selby (eds) *Education and Climate Change: Living and Learning in Interesting Times*, Routledge, London, pp. 13–34.

Henderson, J & Parkes, P 2014, 'Do methods of assessment accurately reflect the priorities of conservation training?' in J Bridgland (ed) *ICOM-CC 17th Triennial Conference Preprints, Melbourne Australia 17–19 September 2014*, ICOM-CC, Paris, pp. 1–11.

James, R & Baldwin, G 2002, *Nine Principles Guiding Teaching and Learning*, Centre for the Study of Higher Education, The University of Melbourne.

Kagawa, F 2009, 'Learning in emergencies: defense of humanity for a livable world' in F Kagawa & D Selby (eds) *Education and Climate Change: Living and Learning in Interesting Times*, Routledge, London, pp. 106–124.

Klein, N 2007, *The Shock Doctrine: The Rise of Disaster Capitalism*, Metropolitan Books, New York.

Kreber, C 2009, 'Supporting student learning in the context of diversity, complexity and uncertainty' in C Kreber (ed) *The University and Its Disciplines*, Routledge, New York and London, pp. 3–18.

Leijonhufvud, G 2021, *Climatic and Environmental Threats to Cultural Heritage: Education* [survey], Project Reference No: 2021-20655-18206-3, University of Melbourne.

Miller, T 2005, 'Foreword' in Plan International 2005, *Children and the Tsunami: Engaging with Children in Disaster Response, Recovery and Risk Reduction*, Plan Ltd, Thailand, p. 2.

Orr, DW 1994, *Earth in Mind: On Education, Environment and the Human Prospect*, Island Press, Washington, DC.

Ploner, J 2011, *Promoting Students' 'Resilient Thinking' in Diverse Higher Education Environments*, Technical Report, The Higher Education Academy – Centre for Sociology, Anthropology, Politics (C-SAP), UK.

Scott, M 2008, 'Conservation interdisciplinarity and pedagogical implications' in J Bridgland (ed) *Preprints of the ICOM-CC 15th Triennial Conference New Delhi India 22–26 September 2008*, ICOM-CC, Paris, p. 122–128.

Scott, M 2016, *Tradition and Innovation: Building the Conservation Discipline in Australia*, PhD Thesis, The University of Melbourne, Australia.

Smithen, P 2021, *Climatic and Environmental Threats to Cultural Heritage: Education* [survey], Project Reference No: 2021-20655-18206-3, University of Melbourne.

Thompson, K 2021, *Climatic and Environmental Threats to Cultural Heritage: Education* [survey], Project Reference No: 2021-20655-18206-3, University of Melbourne.

Tse, N 2021, *Climatic and Environmental Threats to Cultural Heritage: Education* [survey], Project Reference No: 2021-20655-18206-3, University of Melbourne.

Tyler, L 2021, *Climatic and Environmental Threats to Cultural Heritage: Education* [survey], Project Reference No: 2021-20655-18206-3, University of Melbourne.

University of Melbourne (UoM) 2020, 'Sustainability survey 2020 final report', https://sustainablecampus.unimelb.edu.au/__data/assets/pdf_file/0003/3590031/2020-Sustainability-Survey-Report.pdf
Wharton, G 2021, *Climatic and Environmental Threats to Cultural Heritage: Education* [survey], Project Reference No: 2021-20655-18206-3, University of Melbourne.
Widodo, J 2021, *Climatic and Environmental Threats to Cultural Heritage: Education* [survey], Project Reference No: 2021-20655-18206-3, University of Melbourne.

EPILOGUE

Climate change is much more than an environmental crisis. It is an ecological crisis with the potential to destroy one of the most potent tools that people have in the fight to save the planet: the knowledge of the past. The health of this ecology of knowledge is as critical as species health or the health of the environment. People build, maintain and bequeath knowledge between generations through preserving their heritage. Climate change provides new and catastrophic risks to their ability to do so.

This book explores the ecology of knowledge needed in the battle for the planet's future. It does so through the prism of heritage conservation in its broadest sense: the practices, memories, stories and knowledge that are grouped together under the title 'intangible heritage'; the cultural, historic and scientific records that are held in museums, galleries, archives and libraries; and the sites and places that constitute the built environment and the places assigned cultural significance. In doing so, it examines how heritage fits within a broader ecological approach that is cultural, epistemological and environmental. The book also argues that, as humans struggle to recalibrate a world it has so significantly disrupted, it is the diverse nature of this heritage that is important. For this reason, the focus has been on areas understood to be under-represented in the dominant informing documents that support current climate change responses.

This need to bring more diverse voices into climate crisis advocacy is also a reason the book opened with an explanation of narratives as powerful tools in heritage preservation. Policies, strategies, research papers and other documents are a necessary part of understanding and responding to the climate crisis; but they are not the whole story. Knowledge that is held locally, is passed on in language that is threatened, or is not available in requisite formats is often excluded from global domains such as decision-making forums and influential policy documents. This in turn has cascading effects on the ability of heritage experts, including heritage professionals and community leaders, to wield influence in critical forums such as the IPCC. While geographers, geologists, economists, political scientists and others are strongly represented in the researchers who are contributing to climate crisis research and documentation, heritage experts are more likely to be included as invitees who present submissions or represent constituencies. This invited status can marginalise important knowledge.

Part of this problem of representation is recognised as a tension throughout the book, as it grapples with the problem of 'heritage' as an umbrella term to describe sets of activities and knowledge domains that are differentiated by their focus on the format of heritage as place, or collection, or object type (archive, oral history, art, museum object). This is a problem in the face of a global threat as heritage, as an ecology of knowledge, cannot operate effectively as such when it is defined in these silos by its difference. It is also a tension throughout the book because it is irrelevant to the way that many societies understand their heritage. This points to another tension, that of cross-cultural inclusion. The opportunity for Senior Indigenous Knowledge holders to engage in international forums where their knowledge and expertise would be highly relevant is hampered by many factors including language barriers and the lack of culturally appropriate frameworks. A number of examples in the book point to these tensions and identify the need to create more pluralistic platforms for discussions relating to heritage and climate change.

There should be no misconceptions about the scale and complexity of the work required. Heritage, in all its manifestations, is a complex area, mapped by time, place, culture, ontologies, politics, economics, language and much more. An important first step is, therefore, to identify key gaps where research and action are required. Some of these have been highlighted in this book. First, the research into climate change needs to include more diverse sources, particularly from sectors that are under-represented. At present, the cultural heritage sector, with the term used in its broadest sense, has a limited understanding of what is needed to preserve cultural heritage in the face of climate change. As a result, responses are patchy, territorially focussed by geography and discipline and so fail to produce the co-ordinated leadership required. Second, just as the IPCC has managed to bring together a range of diverse disciplines to produce world-leading research, the heritage sector needs to find points of similarity and association. The reliance of a differentiated model of heritage, split by 'tangibility' and 'intangibility', and further distinguished as place-based, built or collected object further fragments research, responses and impact. Third, there is the need for a more consolidated approach to research, which includes data collection, ingestion and sharing. Where co-ordination does occur, this needs to be improved across the various heritage sectors, rather than what we currently see as a siloing of heritage by place, or method of collecting, or format. Current research has the potential to result in an uneven bias in the data, where some findings are privileged over others. This has flow-on effects that can have multiple impacts, including that our understanding of risks across the broader heritage sector is variable and that resources are unevenly distributed. These then have further flow-on effects that can lead to reduced opportunities for advocacy and for a voice at the tables of local and global policy and strategy makers.

Humankind has engineered its way into a crisis of unimaginable proportions. To be able to respond adequately requires culturally diverse, broadly inter-disciplinary and rigorously wrought interconnections that can support the kinds of rapid responses that are needed to deal with a crisis that is already out of control.

This book attempts to respond to this challenge through a multi-disciplinary analysis of key issues, and an assessment of various interdisciplinary methodologies that have potential as effective tools in considering the preservation of heritage in climate crises and emergencies. The stories told here, and the people who tell them, are all engaged in ensuring that heritage has a future. It's time to support them through advancing a compelling story about the role of heritage in the global fight to stem the climate crisis.

INDEX

Note: *Italicized* page numbers refer to figures, **bold** page numbers refer to tables